AF357335

Plant Virus Emergence

Plant Virus Emergence

Editors

Michael Goodin
Jeanmarie Verchot

MDPI • Basel • Beijing • Wuhan • Barcelona • Belgrade • Manchester • Tokyo • Cluj • Tianjin

Editors
Michael Goodin
University of Kentucky
USA

Jeanmarie Verchot
Texas A&M University
USA

Editorial Office
MDPI
St. Alban-Anlage 66
4052 Basel, Switzerland

This is a reprint of articles from the Special Issue published online in the open access journal *Viruses* (ISSN 1999-4915) (available at: https://www.mdpi.com/journal/viruses/special_issues/emergence).

For citation purposes, cite each article independently as indicated on the article page online and as indicated below:

LastName, A.A.; LastName, B.B.; LastName, C.C. Article Title. *Journal Name* **Year**, *Volume Number*, Page Range.

ISBN 978-3-0365-0596-1 (Hbk)
ISBN 978-3-0365-0597-8 (PDF)

Cover image courtesy of Michael Goodin and Jeanmarie Verchot.

Contents

About the Editors

Michael Goodin is a distinguished plant virologist who was a Professor of Plant Pathology at the University of Kentucky, in Lexington, Kentucky. Michael was born in Jamaica. He was an undergraduate at Brock University in Canada, where he received degrees in both Biology and Chemistry in 1989. He received his M.Sc. and then Ph.D. at Pennsylvania State University, where he trained with C. P. Romaine in plant pathology. He was a postdoctoral fellow with Andy Jackson at the University of California, Berkeley, from 1996 to 2002 and has been a faculty member in the Department of Plant Pathology at the University of Kentucky since 2002. His strong research record includes the development of tools for the in planta study of plant–virus interactions that are now used worldwide. He contributed significantly to the generation of the first functional mini-replicon system for a negative-strand plant RNA virus, the sonchus yellow net virus. Michael pioneered the discovery of emerging viruses infecting coffee plants in Brazil. In 2018, he was awarded the title of the American Society for Microbiology Honorary Diversity Lecturer in recognition of his scholarship, research, and creativity.

Jeanmarie Verchot is a Professor in the Department of Plant Pathology & Microbiology and a Fellow of the Institute for Plant Genomics and Biotechnology at Texas A&M University. Jeanmarie was born in New Jersey, USA. She was an undergraduate at Rutgers University, Cook College, in New Jersey and received a degree in genetics in 1987. She received her Ph.D. from Texas A&M University, where she trained with James C. Carrington in microbiology. She was a postdoctoral fellow with David Baulcombe at the Sainsbury Laboratory in the UK from 1996 to 1998. She was a faculty member at Oklahoma State University from 1998 to 2017. She was a Director for the Texas A&M Agrilife Center in Dallas from 2017–2019, then became a fellow of the Institute for Plant Genomics and Biotechnology at Texas A&M University. She is a Professor of Plant Pathology at Texas A&M University. Her strong research record includes uncovering mechanisms for potato virus X cell-to-cell movement, describing the mechanisms for plasmodiophorid vectors to transmit viruses to plant roots, the discovery of new virus genomes in canna lily, the development of a reverse genetic system for rose rosette virus, and detailing the mechanism for potexvirus interaction with the endomembrane system and the role of the unfolded protein response in monitoring potyvirus and potexvirus infection.

Editorial

Introduction to Special Issue of Plant Virus Emergence

Michael Goodin [1,*,†] **and Jeanmarie Verchot** [2,*]

1   Plant Pathology Department, University of Kentucky, Lexington, KY 40546, USA
2   Department of Plant Pathology and Microbiology, Texas A&M University, College Station, TX 77802, USA
*   Correspondence: mgoodin@uky.edu (M.G.); jm.verchot@tamu.edu (J.V.); Tel.: +1-979-845-1788 (J.V.)
†   Passed away.

Academic Editor: K. Andrew White
Received: 24 December 2020; Accepted: 30 December 2020; Published: 1 January 2021

We are pleased to present in this Special Issue a series of reviews and research studies on the topic of *"Plant Virus Emergence"*. The issue includes a series of articles that elaborate on important plant virus diseases that are among the most recent epidemiological concerns. This Special Issue is predicated on the paradigm that plant virus epidemiology, outbreaks, epidemics, and pandemics parallel zoonotic viruses, and can be consequential to global food security [1]. There is evidence that local, regional, national, and global trade of agricultural products has aided the global dispersal of plant virus diseases. Expanding farmlands into pristine natural areas has created opportunities for viruses in native landscapes to invade crops, while the movement of food and food products disseminates viruses creating epidemics or pandemics [2,3]. As covered in several of the submissions, a number of factors drive plant virus emergence. Included in these are direct anthropomorphic activities, reviewed in *"Homo sapiens: the superspreader of plant viral diseases"* [4]. This theme of distribution of cultivated plants around the world, dispersed from their centers of domestication, implicates humans as being responsible for the novel encounters between plants and their pests. Alternatively, in the article *"Disease Pandemics and Major Epidemics Arising from New Encounters between Indigenous Viruses and Introduced Crops Viruses"* [5], Jones considers the phenomenon of spillover, the movement of viruses from non-cultivated vegetation into introduced crop plants. Such a phenomenon parallels that which occurred in the emergence of Ebola [6], Hendra, and Nipah viruses [7]. Climate change has also created favorable environments for insects that vector plant viruses. Regulatory agencies around the globe are working together to prevent or mitigate the introduction of plant viruses into new areas where they can cause devastation to food security. Moreover, plant virus outbreaks not only directly impact food supply, they also incidentally affect human health.

Other articles in this Special Issue describe plant viruses that are studied for their associations with "outbreak" phenomena (sudden appearance), reservoir hosts and "spillover" phenomena (transfer from wild to domesticated areas), common genetics, and reliance on "vectors" for viral spread. In *"Potato virus Y emergence and evolution from the Andes of South America to become a major destructive pathogen of potato and other solanaceous crops worldwide"*, Torrance and Talianksly [8] discuss how PVY emerged from the Andes of South America to become a worldwide threat to potato and other solanaceous crops. Due to its broad host range and recombination between strains, PVY and related potyviruses show an episodic emergence of new strains with new virulences. Torrance and Talianksly make the case for research that integrates phylogenetic analyses and high-throughput next-generation sequencing for the detection, identification, and surveillance of new virus strains. Along these lines, Herath et al. present, in *"Family Level Phylogenies Reveal Relationships of Plant Viruses within the Order Bunyavirales"* [9], the importance of phylogenetics in consideration of emergence negative-sense segmented RNA viruses infecting arthropods, protozoans, plants, animals, and humans. Comprehensive phylogenetic analyses of the four "hallmark" genes demonstrated the relatedness of virus species across host eukaryotic species, herein reinforcing the similar emergence phenomena that occur in both plant and animal systems. Furthermore, for these viruses, the link between plants and animals is made by the vectors of the plant-infecting viruses.

Insects are significant drivers of virus emergence, as featured in submissions by Pinheiro-Lima et al., "*Transmission of the Bean-Associated Cytorhabdovirus by the Whitefly Bemisia tabaci MEAM1*" [10], and Schoeny, et al., "*Can Winged Aphid Abundance Be a Predictor of Cucurbit Aphid-Borne Yellows Virus Epidemics in Melon Crop?*" [11]. These studies confirm that we must avoid dogmatic perspectives on plant virus transmission, as the first demonstration of a whitefly-transmitted rhabdovirus. Heretofore unknown virus–vector combinations should be anticipated as drivers of future outbreaks, as demonstrated for *Bemisia tabaci* Middle East-Asia Minor 1 (MEAM1) transmission of bean-associated cytorhabdovirus to common bean, and with a lower efficiency to cowpea and soybean. However, not only is the particular vector important, but the timing of its occurrence in fields relative to crop age can influence emergence. For example, the abundance of *A. gossypii* during the first two weeks after planting is a good predictor of disease caused by CABYV. Early control of the vector is necessary to minimize the potential for CABYV epidemics in melon crops.

Increasingly, scientists depend upon advanced DNA and RNA sequencing technologies to monitor the emergence and spread of new plant virus strains or species, facilitate novel virus discovery, and uncover the etiology of complex diseases. In the article by Weiland et al., "*RNAseq Analysis of Rhizomania-Infected Sugar Beet Provides the First Genome Sequence of Beet Necrotic Yellow Vein Virus from the USA and Identifies a Novel Alphanecrovirus and Putative Satellite Viruses*" [12], advanced sequencing technology was used to discover a complex of viruses underlying rhizomania of sugarbeet. Historically, BNYVV is the acknowledged primary cause of rhizomania however, this work demonstrates the impact on disease severity of co-infecting viruses like beet soil-borne mosaic virus (BSBMV), beet soil-borne virus (BSBV), beet black scorch virus (BBSV), and beet virus Q (BVQ), and a novel Alphanecrovirus.

SARS-CoV-2 is the third coronavirus to emerge in recent years, SARS and MERS being the other two, suggesting that virus surveillance and discovery in reservoir species is important preparatory work for modeling disease outbreaks and planning for vaccines before their emergence in humans. The discovery of emergent virus species and strains requires further molecular characterization. Moodley et al., in "*Emergence and Full Genome Analysis of Tomato Torrado Virus in South Africa*" [13], report new virus genome sequences for ToTV. Wieczorek et al., in "*Development of a New Tomato Torrado Virus-Based Vector Tagged with GFP for Monitoring Virus Movement in Plants*" [14] produce infectious clones needed to study virus infection and molecular interactions with the hosts. Such infectious clone technology is foundational in understanding virus molecular functions, as well as in developing plant viral-based protein expression vectors. It is in this arena that scientists employ plant-based protein expression systems for plant-based vaccine and pharmaceutical production. Developing plant-made antivirals and vaccines is essential to contain and mitigate outbreaks at their earliest outset, thus mitigating recurrences of the tragic events related to the SARS-CoV-2/COVID-19 pandemic. LeBlanc et al. [15] review the state-of-the-art systems for producing "mammalian-compatible" biomolecules in plants, particularly as related to glycosylation.

We hope that you find the collection in this Special Issue informative and of interest. As the public becomes informed of scientific theories on virus disease emergence and spread during the COVID-19 pandemic, it is timely that plant viruses are included in the discussion. We hope that the articles in this Special Issue on "*Plant Virus Emergence*" highlight elaborate efforts in plant virology that support broad models for virus outbreaks, spillovers, genetics, reliance on "vectors" and human trade for spread, as well as the maintenance of viruses in "reservoir" hosts. In closing, we thank all of the authors for their enlightening contributions to this Special Issue.

Funding: This work was supported by a grant from NSF (IOS #1759034).

Institutional Review Board Statement: Not applicable.

Informed Consent Statement: Not applicable.

Data Availability Statement: Not applicable.

Conflicts of Interest: The authors declare no conflict of interest.

References

1. Jones, R.A.C.; Naidu, R.A. Global Dimensions of Plant Virus Diseases: Current Status and Future Perspectives. *Annu. Rev. Virol.* **2019**, *6*, 387–409. [CrossRef] [PubMed]
2. García-Arenal, F.; Zerbini, F.M. Life on the Edge: Geminiviruses at the Interface Between Crops and Wild Plant Hosts. *Annu. Rev. Virol.* **2019**, *6*, 411–433. [CrossRef] [PubMed]
3. Rojas, M.J.; Macedo, M.A.; Maliano, M.R.; Soto-Aguilar, M.; Souza, J.O.; Briddon, R.W.; Kenyon, L.; Bustamante, R.F.R.; Zerbini, F.M.; Adkins, S.; et al. World Management of Geminiviruses. *Annu. Rev. Phytopathol.* **2018**, *56*, 637–677. [CrossRef] [PubMed]
4. Ranawaka, B.; Hayashi, S.; Waterhouse, P.M.; De Felippes, F.F. Homo Sapiens: The Superspreader of Plant Viral Diseases. *Viruses* **2020**, *12*, 1462. [CrossRef] [PubMed]
5. Jones, R.A.C. Disease Pandemics and Major Epidemics Arising from New Encounters between Indigenous Viruses and Intro-duced Crops. *Viruses* **2020**, *12*, 1388. [CrossRef] [PubMed]
6. Boisen, M.; Hartnett, J.; Goba, A.; Vandi, M.; Grant, D.; Schieffelin, J.; Garry, R.F.; Branco, L.M. Epidemiology and Management of the 2013–16 West African Ebola Outbreak. *Annu. Rev. Virol.* **2016**, *3*, 147–171. [CrossRef] [PubMed]
7. Amaya, M.; Broder, C.C. Vaccines to Emerging Viruses: Nipah and Hendra. *Annu. Rev. Virol.* **2020**, *7*, 447–473. [CrossRef] [PubMed]
8. Torrance, L.; Talianksy, M.E. Potato Virus Y Emergence and Evolution from the Andes of South America to Become a Major Destructive Pathogen of Potato and Other Solanaceous Crops Worldwide. *Viruses* **2020**, *12*, 1430. [CrossRef] [PubMed]
9. Herath, V.; Romay, G.; Urrutia, C.D.; Verchot, J. Family Level Phylogenies Reveal Relationships of Plant Viruses within the Or-der Bunyavirales. *Viruses* **2020**, *12*, 1010. [CrossRef] [PubMed]
10. Pinheiro-Lima, B.; Carvalho, R.D.C.P.; Alves-Freitas, D.M.T.; Kitajima, E.W.; Vidal, A.H.; Lacorte, C.; Godinho, M.T.; Fontenele, R.S.; Faria, J.C.; Abreu, E.F.; et al. Transmission of the Bean-Associated Cytorhabdovirus by the Whitefly *Bemisia tabaci* MEAM1. *Viruses* **2020**, *12*, 1028. [CrossRef] [PubMed]
11. Schoeny, A.; Rimbaud, L.; Gognalons, P.; Girardot, G.; Millot, P.; Nozeran, K.; Wipf-Scheibel, C.; Lecoq, H. Can Winged Aphid Abundance Be a Predictor of Cucurbit Aphid-Borne Yellows Virus Epidemics in Melon Crop? *Viruses* **2020**, *12*, 911. [CrossRef] [PubMed]
12. Weiland, J.J.; Sharma Poudel, R.; Flobinus, A.; Cook, D.E.; Secor, G.A.; Bolton, M.D. RNAseq Analysis of Rhizomania-Infected Sugar Beet Provides the First Genome Sequence of Beet Necrotic Yellow Vein Virus from the USA and Identifies a Novel Al-phanecrovirus and Putative Satellite Viruses. *Viruses* **2020**, *12*, 626. [CrossRef] [PubMed]
13. Moodley, V.; Gubba, A.; Mafongoya, P. Emergence and Full Genome Analysis of Tomato Torrado Virus in South Africa. *Viruses* **2020**, *12*, 1167. [CrossRef] [PubMed]
14. Wieczorek, P.; Budziszewska, M.; Frąckowiak, P.; Obrępalska-Stęplowska, A. Development of a New Tomato Torrado Virus-Based Vector Tagged with GFP for Monitoring Virus Movement in Plants. *Viruses* **2020**, *12*, 1195. [CrossRef] [PubMed]
15. Leblanc, Z.; Waterhouse, P.; Bally, J. Plant-Based Vaccines: The Way Ahead? *Viruses* **2020**, *13*, 5. [CrossRef] [PubMed]

Publisher's Note: MDPI stays neutral with regard to jurisdictional claims in published maps and institutional affiliations.

Editorial

In Tribute to Michael Goodin

Jeanmarie Verchot [1,*], Andrew O. Jackson [2] and Anne E. Simon [3]

[1] Department of Plant Pathology and Microbiology, Texas A&M University, College Station, TX 77802, USA
[2] Department of Plant and Microbiology, University of California, Berkeley, CA 94720, USA; andyoj@berkeley.edu
[3] Cell Biology and Molecular Genetics, University of Maryland, College Park, MD 20742, USA; simona@umd.edu
* Correspondence: jeanmarie.verchot@exchange.tamu.edu

Academic Editor: K. Andrew White
Received: 4 January 2021; Accepted: 6 January 2021; Published: 8 January 2021

It is with great sadness and sympathy for his family and the plant virology community that we convey the passing of Michael Goodin unexpectedly in December 2020. Michael contributed enormously as co-editor of this special issue on *Emerging Plant Viruses* for MDPI-Viruses; hence, we wish to dedicate this issue to his memory. We are calling this issue a passion project in memory of Michael, because, if you ever had the honor to work with him or interact with him at a study section or conference, you know of his exuberance and passion for plant virology. Michael's well-respected research focused mainly on plant-infecting rhabdoviruses and the biochemistry and cell biology of virus/host interactions during infection. His love for coffee led to his recent work with coffee ringspot viruses, which are emerging threats to coffee production and quality. But Michael was much more than an accomplished scientist. He had a special passion for his family, friends, and colleagues. He was passionate about photography and public and science education. He enjoyed travel, particularly in national and state parks, as well as abroad, as illustrated by his outstanding photographs. Michael loved to talk about growing up in Jamaica and the deliciousness of Jamaican cuisine; he even brought goat to cook an unforgettable, authentic Caribbean meal after giving an invited seminar at the University of Maryland!

Dr. Michael Goodin (1967–2020)

Michael obtained a PhD in 1995 from Pennsylvania State University in Pete Romaine's lab. There, he generated a productive thesis that investigated the virus-induced La France disease in cultivated mushrooms. He then spent five years working as a postdoc in Andy Jackson's lab at UC-Berkeley. Michael obtained a faculty position in the Plant Pathology Department at the University of Kentucky in 2002, and was a professor and the Director of the Plant Science Biological Imaging Facility.

In summary, Michael was a wonderful friend and colleague to so many, and an enthusiastic supporter of plant virology. He was vibrant, creative, and a passionate lover of life. We hope that when you read this special issue, and the letter from Michael, you share with us the memory of his desire to spread the word about emerging viruses that can impact the foods we eat (and drink). We hope that those of you who knew Michael Goodin will reflect on his research contributions to plant virology, the sense of community he instilled amongst us, and his inspiration to students and colleagues, and smile when remembering his laughter and joy at conferences.

Thank you Michael, for all that you have done. You will be greatly missed.

Conflicts of Interest: The authors declare no conflict of interest.

Publisher's Note: MDPI stays neutral with regard to jurisdictional claims in published maps and institutional affiliations.

Review

Disease Pandemics and Major Epidemics Arising From New Encounters between Indigenous Viruses and Introduced Crops

Roger A. C. Jones

Institute of Agriculture, University of Western Australia, 35 Stirling Highway, Crawley, WA 6009, Australia; roger.jones@uwa.edu.au

Academic Editors: Michael Goodin and Jeanmarie Verchot
Received: 31 October 2020; Accepted: 1 December 2020; Published: 4 December 2020

Abstract: Virus disease pandemics and epidemics that occur in the world's staple food crops pose a major threat to global food security, especially in developing countries with tropical or subtropical climates. Moreover, this threat is escalating rapidly due to increasing difficulties in controlling virus diseases as climate change accelerates and the need to feed the burgeoning global population escalates. One of the main causes of these pandemics and epidemics is the introduction to a new continent of food crops domesticated elsewhere, and their subsequent invasion by damaging virus diseases they never encountered before. This review focusses on providing historical and up-to-date information about pandemics and major epidemics initiated by spillover of indigenous viruses from infected alternative hosts into introduced crops. This spillover requires new encounters at the managed and natural vegetation interface. The principal virus disease pandemic examples described are two (cassava mosaic, cassava brown streak) that threaten food security in sub-Saharan Africa (SSA), and one (tomato yellow leaf curl) doing so globally. A further example describes a virus disease pandemic threatening a major plantation crop producing a vital food export for West Africa (cacao swollen shoot). Also described are two examples of major virus disease epidemics that threaten SSA's food security (rice yellow mottle, groundnut rosette). In addition, brief accounts are provided of two major maize virus disease epidemics (maize streak in SSA, maize rough dwarf in Mediterranean and Middle Eastern regions), a major rice disease epidemic (rice hoja blanca in the Americas), and damaging tomato tospovirus and begomovirus disease epidemics of tomato that impair food security in different world regions. For each pandemic or major epidemic, the factors involved in driving its initial emergence, and its subsequent increase in importance and geographical distribution, are explained. Finally, clarification is provided over what needs to be done globally to achieve effective management of severe virus disease pandemics and epidemics initiated by spillover events.

Keywords: pandemics; epidemics; global; disease; threat; food insecurity; crop losses; crop failure; indigenous viruses; introduced crops; new encounter; spillover; developing countries; domestication centers; sub–Saharan Africa

---

1. Introduction

Virus disease epidemics and pandemics threaten all types of cultivated plants including those grown to feed the world's human population and its domestic animals, and others grown for ornamental, fiber or medicinal uses [1–7]. Virus epidemics also threaten wild plant communities growing in natural ecosystems [8–13]. With crop plants, they diminish the growth and vigor of infected plants, decrease gross yields and disfigure plant produce. The losses they cause vary from total crop failure to smaller scale, occur worldwide and have an estimated economic global impact of >US$30 billion annually [1,2,6,7,14–18]. They occur in all types of crop plants. These include staple food crops of crucial significance for achieving food security in subtropical and tropical regions [1,4,5,7,19–25]. With mixed species-managed pastures and wild plant communities in natural ecosystems, their detrimental effects on the growth and vigor of infected plants alter plant species composition. In managed pastures, they diminish the proportion of pasture plants versus weeds causing pasture deterioration and an inadequate feed base for livestock [26–33]. In wild plant communities, they alter the species balance and decrease species diversity, which damages ecosystems and can cause genetic erosion potentially leading to species extinction [12,13,34–37].

Development of damaging virus epidemics is favored by the introduction of new crops to parts of the world where they have never been grown before and the adoption of intensive cropping systems both of which lead to new encounters with virulent viruses infecting crops or indigenous vegetation. They are also favored by introduction of vulnerable new cultivars bred for increased yields [1,2,4,19,20,38–40]. In mixed species-managed pastures, damaging virus epidemics are favored by factors such as relative grazing pressure and trampling by domestic animals resulting in increased insect vector numbers and virus spread by vectors or contact transmission [31–33]. In wild plant communities, they are aggravated by factors such as fragmentation into small patches of vegetation enclosed by crops or urban areas, livestock grazing and human disturbance, e.g., woodcutting and flower collection [4,9,10,37,41].

Several of the world's plant virus disease pandemics and major epidemics have resulted from infection with emerging viruses that arose from new encounter situations in which indigenous viruses spread by spillover (= host species jumps) from infected indigenous plants to infect introduced cultivated plants [1,4,5,7,19,20,42]. However, epidemics can also take place when introduced viruses spread to indigenous plants from infected introduced cultivated plants [4,9–13,37]. Thus, on the one hand, when introduced cultivated plants domesticated elsewhere grow next to indigenous wild plants or locally domesticated crop plants they never encountered previously, indigenous viruses associated with these indigenous hosts can spillover to the introduced crop plants causing virus disease epidemics in them. On the other hand, introduced viruses can also spread to indigenous crop or wild plants from infected introduced cultivated plants or associated weeds, causing virus epidemics. Both types of invasions require virus spread to occur at the interface between indigenous and introduced plants [1,4,9–13,37,40–44].

Pandemics or epidemics occurring in diverse crops and all continents, apart from Antarctica, were documented in a series of reviews written by the late Professor Michael Thresh [45]. These reviews covered the period from the inception of plant virology in the early 1900s up to 2006 [1,2,19,20,38,40,46–50]. In 1980, Thresh [1] provided a review of the origins and epidemiology of a wide range of important plant virus diseases. More up-to-date accounts of damaging pandemics or major epidemics involving several mostly single virus–host–vector pathosystems were described in several recent reviews [51–58]. In addition, a recent review focused on the global dimensions of plant virus disease [7].

This review describes virus disease pandemics and major epidemics that arose from spillover scenarios involving new encounters between indigenous viruses and introduced crops, rather than virus spread from introduced crops to indigenous crops or natural vegetation. It does this by providing historical and up-to-date information on five examples of virus diseases that threaten staple food crops critically important for food security in developing countries, placing special emphasis on the situation in sub-Saharan Africa (SSA). The sixth virus disease example threatens livelihoods in SSA because it devastates production of a valuable food export crop. In addition, brief coverage is provided of several other examples of major virus disease epidemics that arose from new encounters between indigenous viruses and introduced crops important for food security in different parts of the world.

2. General Concepts

2.1. Definitions

In his 1970 review of 'catastrophic plant diseases', Klinkowski [59] emphasized that many plant disease agents, including viruses, cause epidemics and pandemics, especially when they spread from their centers of origin into continents where they were formerly absent. He defined an epidemic as being "where a disease is spread over an area in which its causal agent has been present for a long time"; a progressive epidemic as "where it expands from this area into others"; and a pandemic as "where epidemics cause mass infections spread over several continents". He gave five plant virus disease examples: sugarcane mosaic disease spreading worldwide fitted his 'pandemic' definition; plum pox, sugar beet yellows and tobacco veinal necrosis diseases spreading mostly in Europe matched his progressive epidemic definition; and cocoa swollen shoot disease (CSSD) spreading in Ghana, West Africa matched his epidemic definition. Subsequently, in plant virology, the term progressive epidemic has fallen into disuse and a plant virus disease pandemic has come to include "an epidemic occurring over a very wide area, crossing international boundaries and causing severe crop losses" [23]. In practice, however, the term epidemic is now widely used to cover all three of these types of epidemic situations, while the term pandemic has become restricted mainly to damaging virus diseases that spread widely between different countries in SSA, e.g., CSSD [18] and cassava mosaic disease (CMD) [23] and cassava brown streak disease (CBSD) [52]. In this review, the 'pandemic' definition now mainly used in Africa is also applied to other continents, otherwise the term 'epidemic' is used.

An emerging virus is usually considered to be "one that causes damaging epidemics but has only evolved or been recognized recently, changed its pathogenesis, increased its host range or increased its geographical distribution" [3,55]. Further, a re-emerging virus is usually considered to be "one that once caused serious disease problems, but then declined in importance before suddenly increasing in incidence and geographical distribution causing considerable crop damage" [4]. Therefore, the term virus emergence refers to "the first appearance of a virus and its associated initial increase in incidence/geographic range", and the term virus re-emergence refers to "the reappearance of virus and its associated increase in incidence/geographic range". When the term vulnerable is applied to a crop cultivar [20], this means that "the cultivar is both susceptible to virus infection (i.e., it becomes infected readily), and sensitive to infection once systemic infection has occurred (i.e., it develops severe symptoms)" [1,60]. Thus, susceptible is the opposite of resistance and sensitive is the opposite of tolerance [60]. The term virus spillover refers to "spread of a virus from naturally-infected host to a new host it has not encountered previously", and the term spillback refers to "spread of a virus from the new host back to the natural host" [42].

2.2. Crop Domestication Centers and Introductions

Selection of local land races of crop plants from wild ancestors commenced more than 10,000 years ago in the worlds' plant domestication centers [61,62]. Viruses from these wild ancestors were present among the land races derived from them and these indigenous viruses adapted to their new situation multiplying in cultivated plants growing mostly in mixed species cultivation [1,4,9]. Later, through international trade, crop plants were moved progressively away from their domestication centers to distant continents where they were often grown as monocultures. For example, the Columbian Exchange was responsible for the introduction of crops critical for food security to other continents following the Spanish 1492 arrival in the Americas, such as maize (*Zea mays*), cassava (*Manihot esculenta*), potato (*Solanum tuberosum*) and tomato (*Solanum lycopersicum*) [63]. In consequence, new encounters between introduced cultivated plants, and infected wild or crop plants occurred resulting in spillover of indigenous viruses into introduced crops. Sometimes epidemics arose soon afterwards and sometimes only after a considerable delay triggered by other factors, and some later became pandemics [1,4,9,18–23,40,52].

2.3. Factors Favoring Spillover

Successful spillover starts with spread of already existing genetic virus variants from a virus infection source plant to the new host plant, and the outcome for each individual variant depends on its relative abilities (i.e., fitness) to survive once it infects each host, adapt to new hosts or vectors and achieve efficient epidemic spread [64]. A range of factors favor successful virus spillover, emergence or re-emergence. These include: presence of efficient indigenous or introduced virus vectors, including "supervectors"; introduction of vulnerable crop cultivars; adoption of cultural practices involving agricultural intensification, extensification and diversification; intensive wildflower production and conservation projects; the relative ability of a virus to generate virulent new variants through mutation, reassortment and recombination; and climate change arising from global warming [1,2,4,16,19,20,42,55,64–73].

3. Rice Yellow Mottle Disease

Asian rice (*Oryza sativa*) is a cereal crop domesticated from wild rice in China approximately 10,000 years ago. It soon spread from there to Southeast Asia, the rest of East Asia and the Indian subcontinent, next to the Middle East, Europe and North Africa, and more recently to the Americas and Oceania. Approximately 1000 years ago, it was introduced to East Africa where it was grown in coastal regions. In the second half of the 19th century, it was taken inland to be sown in the rest of East Africa, Central Africa and then taken to West Africa and Madagascar. The inland delta of the upper Niger River was where African rice (*Oryza glaberrima*) was first domesticated 3000 years ago. It spread gradually from there within West Africa [74,75]. Overall, rice is ranked as third in importance as a staple food crop but in the developing world it is ranked first [76,77]. Many viruses cause disease epidemics in this crop [78]. An example of a major rice virus disease epidemic that arose by virus spillover and now endangers developing country food security is described below.

Rice yellow mottle disease (RYMD) was first described in 1966 infecting rice plantings in the Lake Victoria region of Kenya in East Africa (Table 1). This initial appearance coincided with one of Africa's first intensive irrigated rice production programs. Afterwards, on several different occasions, such programs triggered RYMD appearance in other locations in both East and West Africa. RYMD then spread to most rice-growing countries in other parts of SSA and by 1989 had spread to the island of Madagascar [1,55,75,79,80]. Since the mid-1990s, it has caused a disease epidemic of major economic significance in rice-growing regions and become a major deterrent to rice cultivation in SSA. Both irrigated and rainfed rice develop RYMD but its incidences are higher in irrigated crops [1,55,75,78–81]. However, it has not yet spread elsewhere in the world. RYMD foliage symptoms in rice consist of leaf yellowing, plant stunting, diminished tillering and poor panicle filling, and are associated with low seed production and poor grain quality. The disease causes yield losses of 25–100% [55,78].

Table 1. Nine examples of virus disease pandemics or major epidemics of critical importance for global food security initiated by encounters between indigenous viruses and introduced crops.

Disease	Continents or Regions Currently Affected	Causal Agent(s)	Virus Genus	Vector(s)	Crop Diseased	Crop Origin	Impact	Virus(es) Origin(s)	Causes(s) of Emergence at Interface	Wild Spill over Hosts	Factors favouring Increased Importance/Distribution	Key Citations
Cassava brown streak disease	East, Central and southern Africa	Cassava brown streak and Uganda cassava brown streak viruses	*Ipomovirus*	Whitefly (*Bemisia tabaci*)	Cassava (*Manihot esculenta*)	Amazon rainforest	Widespread devastating yield losses	Coastal East and southern Africa; areas below 1000 m in altitude inland in East and Central Africa	Cassava introduction; spread by whitefly vectors	Wild tree cassava (*Manihot glaziovii*), unknown wild species	Growing vulnerable cultivars; trade in contaminated cassava cuttings; introductions of whitefly supervectors	[illegible]
Cassava mosaic disease	Sub-Saharan Africa and offshore islands, incl. Madagascar; (less affected so far: South India, Sri Lanka and Southeast Asia)	Cassava mosaic virus complex (seven viruses in sub-Saharan Africa/Madagascar. (Two further viruses in South India, Sri Lanka and Southeast Asia)	*Begomovirus*	Whitefly (*Bemisia tabaci*)	Cassava (*Manihot esculenta*)	Amazon rainforest	Widespread devastating yield losses; food shortages acute famine; deaths	Sub-Saharan Africa and Madagascar; East Africa a major center of diversity. (Indian subcontinent and Southeast Asia, separately)	Cassava introduction; spread by whitefly vectors	Several wild *Euphorbiaceae* and *Fabaceae* species	Growing vulnerable cultivars; trade in contaminated cassava cuttings; recombination generating virulent new variants; introductions of whitefly supervectors	[illegible]
Cacao swollen shoot disease	West Africa	Cacao swollen shoot virus	*Badnavirus*	Mealybugs (*Planococcides njalensis* and *P. citri*)	Cacao (*Theobroma cacao*)	Amazon rainforest	Widespread devastating yield losses; infected trees soon killed. Most costly virus eradication program ever.	West Africa	Cacao introduction; spread by mealybug vectors	Tree species *Cola chlamydanta*, (*C. gigantean Adansonia digitata, Ceiba pentandra*, and *Sterculia tragacantha*)	Large-sale planting of vulnerable cacao cv. Amelonado grown as a monoculture	[illegible]
Groundnut rosette disease	Sub-Saharan Africa and offshore islands, including Madagascar	Groundnut rosette virus, groundnut rosette assistor virus and virus satellite	*Umbravirus, Luteovirus*, virus satellite tri-partite complex	Aphid (*Aphis craccivora*)	Peanut (= groundnut), *Arachis hypogea*)	South America (several locations)	Devastating yield losses, crop failure; major deterrent to peanut cultivation	Sub-Saharan Africa	Peanut introduction; spread by aphid vectors	*Physalis peruviana* and *Cassia obtusa*	Cultural practices including late sowing, wide row spacing	[illegible]
Tomato yellow leaf curl disease	All continents except Antarctic	Tomato yellow leaf curl virus. (Several other begomoviruses also cause TYLCD, but have localised distributions—see Section)	*Begomovirus*	Whitefly (*Bemisia tabaci*)	Tomato (*Solanum esculentum*)	Andean region (Peru, Ecuador)	Widespread devastating yield losses, crop failure, hunger	Middle East and Iran	Tomato introduction; spread by efficient whitefly vectors	Several wild tomato species	International trade in tomato seedlings infected with TYLCV's Mld and IL strains and carrying efficient whitefly vector MEAM1 and MED cryptic species; wind currents carrying viruliferous whitely vectors	[illegible]
Maize rough dwarf disease	Mediterranean region and Middle East	Maize rough dwarf virus	*Fijivirus*	Planthopper (*Laodelphax striatelus*)	Maize (*Zea mays*)	Mexico	Devastating yield losses in hybrid maize cultivars; major threat to maize crop	Mediterranean region and Middle East	Maize introduction; spread by planthopper vector	*Digitaria sanguinalis* and other wild grasses	Growing vulnerable hybrid maize cultivars	[illegible]

Table 1. *Cont.*

Disease	Continents or Regions Currently Affected	Causal Agent(s)	Virus Genus	Vector(s)	Crop Diseased	Crop Origin	Impact	Virus(es) Origin(s)	Causes(s) of Emergence at Interface	Wild Spill over Hosts	Factors favouring Increased Importance/Distribution	Key Citations
Maize streak disease	Sub-Saharan Africa	Maize streak mosaic virus	*Mastrevirus*	Leafhoppers (*Cicadulina mbila* and nine other *Cicadulina* species)	Maize (*Zea mays*)	Mexico	Widespread devastating yield losses, famine in some years	Southern Africa	Maize introduction; spread by leafhopper vectors; appearance of recombinant strain MSV-A	*Digitaria*, sp. and other wild grasses	Growing vulnerable short-season hybrid maize cultivars; agricultural intensification	[55,71,97,98]
Rice hoja blanca disease	South, Central and North America	Rice hoja blanca virus	*Tenuivirus*	Planthopper (*Sogatodes orizicola*)	Rice (*Oryza sativa*)	China	Devastating yield losses in some years	South America	Rice introduction; spread by its leafhopper vector	*Echinochloa colona* and other wild grasses	Intensified rice cropping; growing vulnerable rice cultivars; long-distance spread by viruliferous leafhopper vector	[78,99]
Rice yellow mottle disease	East, Central and West Africa, and Madagascar	Rice yellow mottle virus	*Sobemovirus*	Contact, beetles, mammals soil and water	Asian rice (*Oryza sativa*) and African rice (*Oryza glaberrima*)	China	Widespread devastating yield losses; major deterrent to rice cultivation	East and West Africa	Spread by contact and vectors; intensification of rice production; Introduction of Asian rice	Wild rice: *Oryza barthii*, Orryza *longistaminata*. Wild grasses. *Echinochloa colona, Eragrostis atrovirens* and *Panicum repens*	Intensive irrigated rice production. Short-distance spread by beetle vectors, contaminated soil, irrigation water, machinery. Long-distance spread by trade in live rice seedlings, stubble and ratoons	[1,55,75–79]

The causal agent of RYMD is rice yellow mottle virus (RYMV; genus, *Sobemovirus*, family, *Sobemoviridae*). RYMV infection occurs naturally in cultivated African and Asian rice, the wild rice species *O. barthii* and *O. longistaminata*, and the wild grasses *Echinocloa colona*, *Eragrostis atrovirens* and *Panicum repens* [78,79]. RYMV has stable spherical virions that remain infectious for long periods on contaminated surfaces and reach high concentrations in infected plants [78]. It is therefore readily contact transmitted, including by wind-meditated plant-to-plant contact transmission [55,100]. It is also transmitted by several chrysomelid beetle species, its most efficient beetle vector being *Sesselia pussilla*. In addition, it is transmitted by mammals and in irrigation water and soil, but is not seed transmitted to seedlings [1,55,75,78,79,100]. Carry over between cropping periods occurs mainly in infected rice stubble arising from incompletely decomposed contaminated plant debris, allowing crops to regenerate from tillers growing from these stubbles (ratooning) and infected wild hosts [75,78,79].

Up until the 1960s, rice was only grown in small-scale subsistence plantings in SSA. In the 19th century in coastal East Africa, RYMV emerged in Asian rice plantings via virus spillover from nearby wild rice and grass hosts, and then spread inland. In West Africa, at the end of the 19th century, a similar spillover process resulted in its emergence in African rice plantings in the upper Niger River delta region, and its spread elsewhere in this region. In both instances, its emergence was attributed to its spread by contact and vectors to rice, and intensification of rice production at the natural and managed vegetation interface under subsistence farming conditions [75]. The introduction of large-sale, intensive irrigated rice production schemes, including irrigation over much of SSA led to its initial detection in Kenya in 1966, development of a major RYMD epidemic and the resulting widespread severe production losses in most rice-growing SSA countries. Irrigation allowed extensive growth of volunteer cultivated rice, wild rice and weed grass plants that remained present during the dry season providing an infection reservoir for RYMV spread in the following growing season [1,75,79].

What was responsible for the increase in geographical distribution of RYMV infection in rice crops found since 1966 within SSA? Since RYMV is not seed borne, widespread dissemination via the seed trade can be discounted. Although spread by vectors from infected alternative hosts or via contaminated irrigation water, soil containing plant debris or agricultural machinery could account for local spread, but they would not account for its rapid long-distance dissemination. Rakotomalala et al. [81] suggested that the rice trade might have been responsible for spreading RYMV from continental Africa to Madagascar. Thus, unknowingly transporting RYMV-infected live rice seedlings, stubble or ratoons to Madagascar, and planting them there, would have introduced the virus. Such introduction via trade could also explain its spread from one country to another within continental Africa, but direct evidence of what actually occurred is lacking [75,79]. Since rice is ranked as the most important staple food crop in the developing world (see above), spread of RYMV to other rice-growing regions of the world leading to a major global epidemic would constitute a further cause for concern over future food security.

4. Cassava Mosaic Disease

Cassava is a perennial tuberous root crop domesticated 10,000 years ago in the Amazonian rain forest region of South America. It is ranked fifth in global importance as a staple food crop, and is currently the third most important food staple in developing countries [77,101] where it is mainly grown by smallholder farmers [54]. In the 16th century, it was taken to West Africa. By the beginning of the 19th century, it was being grown throughout West, Central and East Africa, and had also been introduced to the Indian subcontinent and Southeast Asia. During the 20th century, its cultivation greatly increased in SSA and southern Asia. Africa is now responsible for more than half of its global production. It is propagated vegetatively and grows well in the world's tropical regions, tolerates poor soils and drought, requires minimal inputs, and delivers a high output of energy per hectare [77,101]. Cassava crops become infected with several virus diseases [82]. Two examples of devastating cassava virus disease pandemics that arose by virus spillover and are now endangering food security in developing countries are described below in this section, and in Section 5.

CMD was first found in 1984 in East Africa. By the 1940s, its presence had been reported in most SSA countries that grow this crop (Table 1). It now occurs in all SSA countries where cassava is grown, and, through trade in contaminated cassava cuttings, has spread to islands adjoining Africa [52,54,55,80]. Up until the early 1980s, attempts to manage CMD were restricted to places where its epidemics threatened rural livelihoods and caused food insecurity. These epidemics occurred in vulnerable cultivars. Unfortunately, such cultivars were generally the ones most preferred by smallholder farmers. This was due to the greater yields of higher quality tuberous roots they produced when harvested from healthy plantings [19,20,54,83]. After a virulent form of CMD that affected vulnerable cultivars very severely emerged in the late 1980s and infected cassava cuttings were planted widely, a highly destructive CMD epidemic arose in Uganda. It caused devastating losses in tuberous root production. Many rural inhabitants suffered an almost complete income loss, food shortages developed and famine-induced deaths occurred [19,20,23,102]. It then spread from Uganda to 10 other countries in East and Central Africa resulting in a disastrous CMD pandemic which caused enormous economic losses often accompanied by acute famine [19,20,23,52,55,83,103]. The foliage symptoms associated with CMD consist of severe leaf mosaic and deformation (Figure 1A), and plant stunting, sometimes resulting in plant death [104]. Up to 85% losses in tuberous root yields develop in CMD-affected plants of sensitive cultivars. However, some less widely grown cultivars are more tolerant, and so suffer smaller yield losses [20,105].

Figure 1. (**A**) Field of cassava devastated by cassava mosaic disease. Remaining upper leaves on diseased, mostly defoliated plants show symptoms consisting of severe mosaic and leaf deformation, image modified from [7]. (**B**) Roots of cassava showing marked constrictions caused by cassava brown streak disease (CBSD) (image credit @Natural Resource Institute/Maruthi Gowda). (**C**) Tuberous roots of cassava cut along their lengths showing dry necrotic rotting caused by CBSD (image credit @Natural Resource Institute/Maruthi Gowda). (**D**) Tuberous roots of cassava cut in cross section of showing dry necrotic rotting caused by CBSD (image credit @Natural Resource Institute/Maruthi Gowda). (**E**) Field of tomato devastated by tomato yellow leaf curl disease (TYLCD). All plants have symptoms of diminished leaf size, bunched growth, plant stunting, and lack of fruit formation, image modified from [5]. (**F**) Tomato plant showing severe symptoms of small, pale and upcurled leaves, bunched growth and plant stunting caused by TYLCD following early growth stage infection.

In 1983, the first CMD causal agent was described, African cassava mosaic virus (ACMV; genus *Begomovirus*, family, *Geminiviridae*). During the period 1983–2012, five further begomoviruses associated with CMD were found in SSA and one in Madagascar. All seven cassava begomoviruses were persistently transmitted by the polyphagous cryptic whitefly complex *Bemisia tabaci* [19,54,82,106]. Further, several recombinant strains derived from these begomoviruses were identified, and several alternative wild cassava begomovirus hosts belonging to the *Euphorbiaceae* or *Fabaceae* were reported in different parts of mainland SSA [54,80,107]. However, none of the seven begomoviruses causing CMD in SSA, or Madagascar, occur in cassava's South American domestication centre [82]. Instead, these CMD causing begomoviruses all emerged in new encounter scenarios by spillover of indigenous begomoviruses spread by whitefly vectors from naturally-infected wild host plants into cassava after this crop was first introduced to different parts of this region. For example, ACMV, EACMV, South African cassava mosaic virus and cassava mosaic Madagascar virus probably emerged in West Africa, East Africa, South Africa and Madagascar, respectively. East Africa may be a major center of cassava begomovirus diversity as four cassava begomoviruses apparently emerged there. Whitefly vectors were responsible for spreading viruses from local infected alternative wild hosts to cassava resulting in cassava begomovirus emergence [54].

Following its invasion by indigenous begomoviruses, a combination of diverse factors was responsible for the development of CMD as a major threat to SSA cassava production. These included widespread planting of vulnerable cassava cultivars, widescale distribution of infected cassava planting material, recombination generating virulent new variants, synergistic interactions resulting from mixed cassava begomovirus infections, and frequent introductions of polyphagous whitefly vector types able to reach super-abundant numbers even above 1000 m above sea level [20,54,106]. What was responsible for the virulent form of CMD that caused the highly destructive CMD pandemic that started in Uganda in the late 1980s? This was caused by recombination between EACMV and ACMV resulting in the highly virulent recombinant called the EAMCV-Uganda variant (EACMV-UG). When co-infection occurred between ACMV and EACMV-UG, a synergistic interaction between the two viruses greatly increased virus titer causing very severe disease symptoms [82,102]. Cassava planting material carrying this mixed infection spread rapidly resulting in the disastrous East and Central African CMD pandemic [19,20,23,103]. Moreover, when a cassava mosaic virus is accompanied by a DNA satellite, infection with both may further magnify CMD-induced losses. This is because satellite presence can not only enhance CMD symptom severity but also overcome CMD resistance locus CMD2 enabling infection to occur in otherwise CMD-resistant cassava cultivars or land races that carry it [108].

In the south of the Indian subcontinent and Sri Lanka, CMD also causes major cassava disease epidemics. The principal cassava begomovirus responsible for the epidemics in central and southern India is Indian cassava mosaic virus, and in Sri Lanka it is Sri Lankan cassava mosaic virus (SLCMV). However, SLCMV is also found in southern India [54,80,82]. In addition, CMD caused by SCLMV is currently emerging as an important disease of cassava in Southeast Asia. It was found first in 2016 in Cambodia, and then spread to Vietnam, Thailand and Southernmost China [84].

5. Cassava Brown Streak Disease

CBSD was recorded first in 1936 infecting cassava crops in coastal Tanzania (Table 1). By 1950, it was found at altitudes below 1000 m in coastal East and southern Africa, and inland in Malawi and Uganda [20,57]. For several decades it was mostly ignored, but this changed in the 1990s when it re-emerged as a major factor causing epidemics that greatly diminished production of unblemished cassava tuberous roots and threatened food security. This occurred first in the East African Lake Victoria region, and next in most countries of East Africa, including at altitudes over 1000 m. By 2010, CBSD had spread widely and was causing a pandemic resulting in devastating losses in cassava production in East and Central Africa. Moreover, the likelihood of its further spread posed a serious risk to West African cassava crops [52,57,83,109]. CBSD causes root constriction (Figure 1B) and a brown-black, necrotic rot of cassava tuberous roots (Figure 1C,D). In addition, CBSD diminishes yields of tuberous roots by up to 70%. Its foliage symptoms consist of chlorotic blotching, mottle and veinal chlorosis of leaves, and brown stem streaking, symptom severity varying between cassava cultivars. These foliage symptoms are often too subtle for farmers to recognize and asymptomatic infection also occurs, so disease presence often goes unnoticed within the growing crop. This leads to infected cuttings being distributed for transplanting and farmers not knowing their cassava crop is affected until after its tuberous roots are harvested [20,52,57,83,110].

The two CBSD causal agents are cassava brown streak virus (CBSV; genus *Ipomovirus*; family, *Potyviridae*) and the closely related Ugandan cassava brown streak virus (UCBSV). The whitefly *B. tabaci* transmits both semi-persistently [57,80,111–114]. The foliage and root symptoms that UCBSV elicits differ from those CBSV causes. UCBSV causes circular chlorotic blotches between leaf veins without any veinal associations, whereas CBSV elicits more severe root necrosis, and feathery chlorosis alongside veins from which chlorotic blotches develop [57]. There is as yet no evidence of recombination between CBSV and UCBSV but potentially synergistic mixed infection occurs commonly with both of them and may elicit more severe symptoms. The only alternative hosts reported are the wild perennial tree cassava (*Manihot glaziovii*) and the non-cassava wild species *Zanha africana* and *Trichodesma zeylanicum* in which CBSV was detected, and the wild cassava species *Manihot carthaginensis* in which both viruses were found. Whether these species act as virus reservoirs for CBSV and UCBSV spread to cassava crops is unknown but seems plausible [57,80,115]. CBSV and UCBSV only occur in Africa, and are indigenous ipomoviruses from tree cassava or wild host species occurring within Africa. They emerged in new encounter scenarios within Coastal East and southern Africa, and in areas below 1000 m in altitude inland within East and Central Africa. This emergence was by spillover of the two indigenous ipomoviruses from naturally-infected wild plants to cassava after it was introduced, whitefly vectors being responsible for spreading them to cassava [57,80,112,116]. The likely reasons for the CBSD pandemic disease threatening food security in East and Central Africa were as follows: (i) inadvertent transportation of CBSD-infected cassava planting material to many new locations; (ii) distribution of vulnerable cultivars likely including CMD-resistant ones that turned out later to be CBSD-susceptible; and (iii) frequent introductions of polyphagous whitefly vectors capable of reaching superabundance at over 1000 m above sea level [20,52,54,57,116].

6. Tomato Yellow Leaf Curl Disease

The most important vegetable crop grown worldwide is tomato. It is important for human nutrition as it provides the human body with vitamins, minerals and plant compounds that bestow health benefits, including antioxidants. Ancestral cultivated tomato was originally confined to the central Andean region of South America (now in Peru and Ecuador), where it was first domesticated from wild tomato species. After spreading north in the Americas in pre-Columbian times, its domestication continued in Mexico. It was taken from there to Europe in the 16th century from where it was later distributed globally [117]. At least 136 virus diseases affect the tomato crop [118]. An example of a tomato virus disease pandemic that arose from virus spillover and is now endangering food security in developing countries worldwide is described below.

Globally, the most economically significant tomato virus disease is tomato yellow leaf curl disease (TYLCD) (Table 1). It occurs in the world's tropical and subtropical regions where its epidemics collectively cause a devastating pandemic, which is often the principal factor limiting tomato production. It was reported first in Israel in the 1930s and has severely damaged tomato crops in Middle Eastern countries since the 1960s. It remained restricted to Middle Eastern and eastern Mediterranean countries until the late 1980s. However, in the three decades that followed it spread west to the Western Mediterranean region, Caribbean islands, Central America, North America, and Venezuela in northern South America; south to West and East Africa, and to Reunion Island and Mauritius in the Indian Ocean; and east to the Arabian peninsula, the Indian subcontinent, Southeast Asia and East Asia; and then in 2006 to Oceania [80,91–93]. TYLCV symptoms in tomato foliage consist of leaf upward curling, yellowing and diminished size, flower abortion and plant stunting (Figure 1E,F). TYLCD epidemics cause dramatic losses due to the greatly decreased number of fruit formed. When early plant infection is widespread, the order of magnitude fruit yield loss can reach 100% causing total crop failure. Since tomato is often a major component of the diet of smallholder farmers in many developing countries, severe TYLCD outbreaks in their crops leads to hunger, indebtedness and farm abandonment [80,91–93].

The causal agent of TYLCD is tomato yellow leaf virus (TYLCV; genus *Begomovirus*, family, *Geminiviridae*) which is persistently transmitted by the whitefly *B. tabaci*. Tomato is its primary host, but it also naturally infects some alternative hosts sporadically, including common bean (*Phaseolus vulgaris*), the solanaceous ornamentals petunia (*Petunia hybrida*) and lisanthus (*Eustoma* spp.), and several wild tomato species [71,80]. TYLCV itself is subdivided into seven distinct strains but only the mild (Mld) and Israel (IL) strains have been dispersed widely outside the Middle East [92]. *B. tabaci* is a polyphagous supervector that exists as a species complex. Its cryptic species MEAM1 (= biotype B) and MED (= biotype Q) are its most efficient transmitters [68,92,119]. In the field, *B. tabaci* transmits TYLCV from infected to healthy plants both locally and, when viruliferous whitefly are blown over greater distances in wind currents. Some strain TYLCV-IL variants may be seed transmitted in tomato [120].

TYLCV is an indigenous virus to the Middle East. Somewhere in between the Jordan Valley eastwards to Iran, it emerged in a new encounter scenario at the interface between natural and managed vegetation. This emergence occurred by spillover from unidentified indigenous TYLCV-infected host sources into the introduced tomato crop. As mentioned above, tomato was domesticated in the Andean Region of South America. Factors favoring its emergence from infected indigenous plant hosts included efficient vector transmission by the cryptic species of the *B. tabaci* complex present and TYLCV's ability to infect tomato crop plants readily [71,121]. Its widespread dissemination and establishment globally has been attributed to international trade in tomato seedlings unknowingly infected with TYLCV and infested with viruliferous *B. tabaci* MEAM1 or MED cryptic species. In addition, inadvertent international trade in TYLCV-infected tomato fruits and seedlings might also have been implicated [4,68,92,121,122].

Several other begomoviruses that infect tomato, and are relatives of TYLCV, cause TYLCD locally in some world regions. This includes tomato yellow leaf curl China virus, tomato leaf curl Malaysia virus; tomato yellow leaf curl Kanchanaburi virus, tomato yellow leaf curl Malaga virus, tomato yellow leaf curl New Delhi virus, tomato yellow leaf curl Sardinia virus, and tomato yellow leaf curl Thailand virus [53,68,80,122–126]. However, TYLCV is generally more invasive than these other tomato begomoviruses, which are mostly restricted to regions in which they are indigenous, so it tends to displace them [127].

7. Groundnut Rosette Disease

Grain legume crops such as peanut (=groundnut; *Arachis hypogaea*) are important for human nutrition and achieving sustainable food production. Their greater use would improve food security considerably not only by improving human and livestock health but also by improving soil fertility through fixing atmospheric nitrogen [128]. Currently, one of the major factors holding back their wider usage is lack of consistently in obtaining high yields regularly due to virus disease epidemics [129–131]. This applies to grain legumes grown in warmer climates such as peanut, common bean, cowpea (*Vigna unguiculata*), mung bean (*Vigna radiata*), and soybean (*Glycine max*), and cool season grain legumes, such as chickpea (*Cicer arietinum*), faba bean (*Vicia faba*), field pea (*Pisum sativum*), lentil (*Lens culinaris*) and lupin (*Lupinus* spp.) [131–133]. An example of a major grain legume virus disease epidemic that arose by virus spillover, and now endangers food security in developing countries is described below.

Peanut is an important crop that helps to ensure developing country food security. More than 6000 years ago, it was domesticated independently in several locations in South America. It was introduced to SSA in the 16th century [134]. The most important virus disease of peanut in SSA is groundnut rosette disease (GRD) [87,88]. It was first reported in Tanzania in 1907, and later elsewhere in SSA and its offshore islands, including Madagascar (Table 1). It causes a destructive disease in many countries of East, West, Central and southern SSA, and Madagascar [19,87–89]. Although generally sporadic in occurrence, GRD epidemics can be very damaging resulting in almost total peanut crop failure, and their unpredictability greatly hampers attempts to manage the disease. In semiarid tropical conditions, they cause yield losses of greater magnitude than any other peanut virus disease. GRD poses a very serious constraint to peanut production, and its epidemics often cripple the rural economy, causing smallholder farmers to abandon growing the crop [19,87–89]. GRD foliage symptoms consist of two main types, 'chlorotic rosette' (chlorotic yellow leaf mosaic and rosette; Figure 2A–C) and 'green rosette' (green leaf mosaic and rosette; Figure 2D). Chlorotic rosette occurs throughout SSA, but green rosette is less widely distributed. Both types of rosette syndromes cause young diseased plants to appear bushy, and become severely stunted. Diseased older plants only develop symptoms in some of their shoots or parts of their shoots. Yield losses are greatest in young plants, and can reach 100% resulting in complete crop failure when widespread infection occurs before flowering time [88].

Figure 2. (**A**) Field of peanut with a large central area of chlorotic severely stunted plants caused by the chlorotic rosette syndrome of groundnut rosette disease (GRD) (image credit @Washington State University/Naidu Rayapati). (**B**) Peanut plant showing chlorotic (yellow) leaf mosaic symptoms caused by the chlorotic rosette syndrome of GRD (image credit @Washington State University/Naidu Rayapati). (**C**) Row of peanut plants showing bushiness and severe plant stunting caused by the chlorotic rosette syndrome of GRD (right and center), healthy plant on left (image credit @Washington state University/Naidu Rayapati). (**D**) Row of peanut plants showing bushiness and severe plant stunting caused by the green rosette syndrome of GRD (right), healthy plants on left (image credit @Washington state University/Naidu Rayapati). (**E**) Cacao tree showing swollen trunk symptom (pointed to by arrow) caused by cacao swollen shoot disease (CSSD) (image credit @International Institute of Tropical Agriculture/Lava Kumar). (**F**) Shoot of cacao tree showing characteristic swollen shoot symptom (pointed to by arrow) caused by CSSD (image credit @International Institute of Tropical Agriculture/Lava Kumar).

GRD is elicited by a tripartite virus complex consisting of groundnut rosette virus (GRV; genus, *Umbravirus*, family, *Tombusviridae*), groundnut rosette assistor virus (GRAV; genus, *Luteovirus*, family, *Luteoviridae*) and satellite RNA (sGRV). *Aphis craccivora* (the cowpea aphid) transmits this tripartite virus complex persistently. It is not seed borne. Presence of GRAV is essential for transmission of GRV and sGRV to occur. Infection with GRAV and GRV but without sGRV fails to elicit any symptoms since sGRV is required for symptom expression [19,87–89]. The legume crops common bean, cowpea, soybean and mung bean, and two weed species, *Physalis peruviana* and *Cassia obtusa* are potential alternative hosts for the tripartite virus complex [90]. Carry-over of infection between growing seasons may occur in infected volunteer peanut plants or infected alternative host species. In addition to these infection sources, overlapping plantings of old infected peanut crops also act as reservoirs of the virus complex for spread to new crops within the growing season. *A. craccivora* vectors spread this complex to the peanut crop [19,87]. The GRAV, GRV and sGRV complex is indigenous to SSA. However, GRAV itself is also present on its own in the Indian subcontinent, Southeast Asia and Oceania [80]. The GRAV, GRV and sGRV complex emerged in SSA by spillover from infected wild vegetation spread by its aphid vector to the peanut crop after this crop was introduced [19,87–89]. Which factors favored development of the GRD epidemics in peanut crops in SSA? Thresh [19] referred to several cultural practices preferred by smallholder farmers that would have contributed to this. These included sowing peanut late in the growing season following the cereal (normally maize) harvest and using wide row spacing to save scarce seed supplies. *A. craccivora* vectors and virus reservoirs were most abundant at this stage of the growing season and wide row spacing attracted incoming aphids to land on peanut plants, both of which favored virus spread. For various reasons, GRD control recommendations to sow earlier using narrow row spacing proved too inconvenient to be adopted by the smallholder farmers.

8. Cacao Swollen Shoot Disease

Cacao (*Theobroma cacao*) is an evergreen, understory tree indigenous to the Amazonian rain forest in South America. It was introduced to West Africa in the second half of the 19th Century where it was mostly planted in lowland forest areas [1]. Its beans are very important to the global confectionery industry as both chocolate and cocoa powder are derived from the cocoa butter extract obtained from them. Cacao plantations therefore provide an important source of income for farmers in developing countries. An example of a cacao virus disease pandemic that arose by virus spillover is described below.

CSSD was reported first in 1936 in Ghana, West Africa, but had been present for many years beforehand causing a widespread tree dieback syndrome (Table 1). Due to its dependence on cocoa as an export crop, by killing millions of trees, the CSSD pandemic caused enormous losses to Ghana's economy, other West Africa countries, such as the Ivory Coast, Nigeria and Togo, experiencing similar losses [18,20]. Indeed, during the period 1946–1997, the Ghanaian eradication campaign against CSSD had cut down 193 million trees [18,20]. By 2020, CSSD has still not been contained effectively, despite this eradication program having being underway for >70 years and constituting the most costly such virus eradication program ever anywhere in the world [85,86]. CSSD's symptoms in infected cacao trees include swelling of the trunk (Figure 2E) and at shoot nodes, internodes (Figure 2F) or tips and on roots, leaf chlorosis and vein banding, and tree dieback [85,86]. CSSD reduces cacao bean yields by 25% in the initial infection year, 50% in the second year and normally within 3–4 years then proceeds to kill cacao trees [135].

CSSD is caused by cacao swollen shoot virus (CSSV; Genus, *Badnavirus*, family, *Caulimoviridae*), which is transmitted semi-persistently by several mealybug species. Its most efficient mealybug vectors are *Planococcoides njalensis* and *P. citri*. Its alternative natural hosts are five indigenous West African tree species, *Adansonia digitata*, *Ceiba pentandra*, *Cola gigantean*, *C. chlamydanta*, and *Sterculia tragacantha*. Its mealybug vectors spread it from infected to healthy trees [85,136,137]. CSSV emerged at the managed and natural vegetation interface in West Africa by spillover from is indigenous tree hosts into cacao trees introduced from Amazonia. In eastern Ghana, this is thought to have involved the native forest understory tree *C. chlamydanta* as the virus source because this tree species is commonly CSSV-infected and colonized by mealybugs, grows in close proximity to cacao plantings and was infected by the same CSSV strain as that found in nearby cacao trees [1]. In other countries, it is unclear whether alternative indigenous CSSV host species other than *C. chlamydanta* growing near cacao plantings were involved in its emergence, as, where this issue was studied, spread by mealybug vectors possibly occurred to them from infected cacao instead of in the opposite direction [136].

Unfortunately, the initial West African CSSV epidemics were exacerbated by another factor as almost all the first large-sale plantings were a monoculture of cacao cv. Amelonado, which had come directly from Amazonia. Although ideal for producing high quality cacao beans and growing well in West African lowland forest areas, this cultivar proved very vulnerable to CSSV infection which rapidly kills it. Although more CSSV-tolerant cacao cultivars with resistance to mealybugs were introduced subsequently, in 2006 there were still large areas of cv. Amelonado plantings in West Africa likely to suffer damaging CSSV epidemics [20]. Thus, given the significance of cacao beans as the only source of the key ingredient for chocolate and its confections, cacao provides an example of a globally important crop being threatened by a highly damaging pandemic resulting from introduction of a new crop to another continent, and being aggravated by large-scale planting of a vulnerable cultivar.

9. Other Virus Diseases

Table 1 provides details of three other major virus disease epidemics arising from spillover of indigenous viruses from infected wild plants into introduced cereal crops of critical importance for global food security. These crops are (i) maize, which is not only the world's most important staple food crop overall, but also the third most important in the developing world [77,138]; and (ii) Asian rice, which, conversely, is not only the developing world's most important staple food crop, but also the third most important overall (see Section 3). Maize was domesticated in Mexico. Following the Spanish conquest of the Americas in the 15th century, it was dispersed from there to other continents reaching Europe and Africa in the 16th and 17th centuries, respectively [139].

Maize streak disease (MSD) and maize rough dwarf disease (MRDD) both arose by new encounters at the managed and natural vegetation interface (Table 1). This was by spillover of the indigenous viruses maize streak virus (MSV, genus *Mastrevirus*, family, *Geminiviridae*) and maize rough dwarf virus (MRDV; genus *Fijivirus*, family, *Reoviridae*) spread by their respective hopper vectors from infected *Digitaria* and other wild grass species. With MSV, its leafhopper vectors *Cicadulina mbila* and nine other *Cicadulina* species, and with MRDV, its planthopper vector *Laodelphax striatellus*. With MSV, this spillover occurred in southern Africa [97,98], but with MRDV in Southern Europe and the Middle East [1,20,94,95]. With MSV, the main trigger for its emergence was recombination between its virus strains resulting in virulent recombinant strain MSV-A, which adapted readily to its new host maize [98]. Another important factor contributing to the development of disastrous MSD epidemics was agricultural intensification, including widespread use of vulnerable short-season maize hybrids enabling two overlapping maize crops to be grown per year. Having two crops per year allowed its leafhopper vectors to spread MSV readily from one crop to the next [98]. With MRDV, its emergence by spillover from wild grasses to maize likely occurred well before the 1940s when it caused devastating epidemics in Italian maize crops sown with recently introduced high yielding American cultivars. These American maize cultivars were much more vulnerable than those grown previously. The same scenario unfolded in maize crops in Israel in the 1950s [1,20]. Currently, MRDD outbreaks caused by MRDV remain a major threat to maize production throughout the Mediterranean region [96].

Rice hoja blanca disease (RHBD) arose in northern South America from a new encounter scenario at the managed and natural vegetation interface. It involved spillover of the indigenous rice hoja blanca virus (RHBV, genus *Tenuivirus, family, Phenuiviridae*) spread by its vector planthopper *Sogatodes orizicola* from RHBV-infected plants of *Echinochloa colona* and other wild grasses to rice (Table 1) [78,99]. This spillover to rice likely occurred well before the 1930s when RHBD was recognized as the cause of major rice virus disease epidemics in Colombia. Within two decades, similar disastrous RHBD epidemics occurred in Venezuela, Panama, Costa Rica, Cuba and Florida, and within a further decade throughout subtropical and tropical regions of Americas. RHBVs rapid widespread dissemination was caused by long-distance flights of viruliferous *S. orizicola* leafhoppers. The devastating crop losses that occurred resulted from widespread use of highly RHBV- and *S. orizicola*-susceptible rice cultivars, and intensive cultural practices, such as continuous rice cropping, which favored RHBV spread [99].

Instead of becoming infected by indigenous virus spillover occurring directly from local virus-infected native plants, introduced crops can also become invaded indirectly by virus spread from infection reservoirs consisting of infected plants belonging to indigenous crops, crops introduced previously or introduced weeds [4]. For example, after papaya's introduction from the Americas to Eurasia, the global papaya ringspot disease pandemic that papaya ringspot virus (PRSV, genus, *Potyvirus*, family, *Potyviridae*) elicited is considered to have started in the Indian subcontinent. Aphid vectors spread the virus from PRSV-infected cucurbit crop plants already growing there to papaya plants growing in nearby plantations of this introduced tree crop. PRSV's host adaptation to papaya was attributed to a mutation that enabled cucurbit-adapted PRSV to infect it readily [140]. Such indirect virus spillover via an intermediate crop host seems more likely to occur with generalist than specialist viruses due to their broader host ranges [4].

There are many other examples of food security being impaired by major plant virus disease epidemics that arose after introduced crops domesticated in one continent were introduced to another where indigenous viruses they had never met before infected them. Such emerging virus disease epidemics often result from infection with viruses in the *Begomovirus, Orthotospovirus* and *Potyvirus* genera [1,2,4,7,19,20,25,71,125]. Moreover, as viruses belonging to these three genera are often generalists, with them it is normally unclear whether indigenous virus spillover occurred directly or indirectly into the introduced crop. An example of this is provided by groundnut bud necrosis virus (GBNV; genus *Orthotospovirus*, family, *Tospoviridae*), which is indigenous to the Indian subcontinent and has a wide natural host range. After introduction of the originally South American crop tomato (see Section 6) to the Indian subcontinent and Southeast Asia, GBNV infected it causing a disastrous major epidemic [141–143]. Whether its spillover into tomato was directly from GBNV infection reservoirs in indigenous native plants, or indirectly via such reservoirs already present in other crops or introduced weeds, is unknown.

Begomovirus disease complexes have caused major epidemics in introduced and local vegetable crops growing in Southeast Asia [126]; the Indian subcontinent [53,144]; the Middle East and Mediterranean region [67,145]; SSA [146]; Northern South America [147–150]; and both Central America and Mexico [151]. The many indigenous begomoviruses that occur locally in tomato outside its domestication center, have infected it by indirect or direct spillover from local infection reservoirs. There is evidence some indigenous begomoviruses that infect tomato in Brazil have infected it by direct spillover from wild plant hosts [71]. However, although alternative non-crop hosts of some of these begomoviruses have been identified [53,71,126], in general, whether the original spillover to tomato events occurred directly from such infected wild hosts or indirectly via other already infected crop plants or introduced weeds remains to be determined. However, the critical role played by recombination and pseudo-recombination in begomovirus adaptation to tomato as a new host is well established, e.g., in Southeast Asia [126], the Indian subcontinent [53] and South America [71]. Expansion and intensification of tomato production, and the introduction of the more efficient *B. tabaci* MEAM1 whitefly vector were critical factors in subsequent tomato begomovirus epidemic development [71,122,126,147,152].

10. Management

Preventing initial spillover events that trigger virus emergence from occurring in a newly introduced crop might be possible initially if small-scale plots can be grown on farms where rigorous hygiene standards are maintained. The new crop would have to be grown in such a way as to avoid any possibility of an indigenous virus spreading into it from potentially virus-infected crop or wild plant alternative hosts, and would need regular inspections and virus testing to identify and destroy any potentially virus-infected plants. However, preventing initial spillover events in this way would be extremely difficult or impossible to achieve in practice once the scale of production increases and the crop is being grown widely in different regions, especially in developing countries.

Once a damaging emerging virus disease pandemic or epidemic initiated by spillover is underway in one region of the world, it is important to prevent, or failing that, minimize, further spread of the virus, or virus complex causing it. This requires measures designed to prevent it from entering, establishing or spreading to, and within, other, regions or continents. Strict biosecurity and plant health measures are needed to achieve this. Such measures include quarantine restrictions applied in the exporting country (pre-border), and at land borders, seaports and airports (border), along with virus eradication and containment programs (post border) [153].

Within regions where a virus disease epidemic is already underway, there is no 'one-size-fits-all' approach towards managing its spread within affected farms and fields. What is required to achieve this is devising integrated virus disease management strategies and tactics that suppress its spread most effectively [7,15,16,19,38,130]. These involve the deployment of appropriate combinations of phytosanitary, cultural, chemical and host resistance measures that target different components of the disease cycle and operate in different ways; biological control measures are sometimes included too, but most suited to protected cropping systems [7,15,16,19]. Such integrated strategies must be adjusted to take into account the scale and nature of the agricultural or horticultural production system involved which may vary from smallholder scale to very large-scale, and also according to climatic conditions, and local ecosystem and societal constraints [7,15,16,19]. Thresh [19,38] described what needs to be done to optimize the effectiveness of integrated disease management in tropical regions, including for most of the devastating virus disease pandemics and epidemics described in this review. Unfortunately, his guidance over adopting an integrated approach was often neglected in the past, especially in SSA. This was due to a tendency to focus on breeding crops for virus resistance and chemical control, whilst neglecting phytosanitary and cultural control measures (CSSD being a notable exception to this because of the widely adopted eradication (phytosanitary) campaign against it). Recently, there are signs this situation is changing, e.g., the widescale development of healthy cassava stock programs as a phytosanitary control measure against CMD and CBSD in SSA [82,83], and the inclusion of some phytosanitary and cultural control measures within the integrated disease management approaches Rojas et al. [119] recommended for geminivirus diseases.

Achieving greater success in managing virus disease pandemics or major epidemics in introduced crops after their initiation by indigenous virus spillover events, requires the strengthening of existing collaborative multidisciplinary research networks developed to address them. Where this is currently absent for a virus disease pandemic currently underway, the creation and fostering of a new collaborative network is warranted. Such multidisciplinary networks require collaboration between developed and developing country researchers, and the participation, amongst others, of specialist plant virologists, entomologists, modelers, agronomists, plant breeders, statisticians, and socioeconomics experts [7]. An example of a collaborative network addressing plant virus disease pandemics initiated by spillover events currently threatening food security in SSA is one tackling CMD and CSSD in cassava [83].

11. Conclusions

Improving the human food supply by introducing food crops domesticated in one continent to another continent, or to another part of the same continent, has exposed a major drawback to this endeavor. This drawback is the unforeseen development of damaging virus disease pandemics or major epidemics that arise by spillover of indigenous viruses into the introduced crops once they become established in their new surroundings. This review provides graphic examples of the enormous crop damage, gross yield and quality losses, and harm to the dependent human population that result. Most of the affected introduced crop examples described were domesticated in the Americas (maize, cassava, peanut, tomato, cacao) and distributed elsewhere in the world as part of the Columbian Exchange, but one of them originated in China (rice), being dispersed from there to other continents. Apart from cacao, these are all staple food crops vitally important for food security in developing countries. Moreover, apart from tomato, all are best suited to growing in regions with tropical or subtropical climates, most of which occur in food insecure parts of the developing world. Tomato not only grows well in such regions but also in warm temperate regions and under protected cropping in cool temperate regions.

Historical and up-to-date descriptions are provided of four examples of virus disease pandemics, and two examples of major epidemics, that arose from spillover scenarios involving new encounters between introduced crops and indigenous viruses spreading from infected natural vegetation. Five of these examples concern virus diseases of staple food crops that threaten food security in developing countries. These examples include those caused by CMD, CBSD, GRD, RYMD in SSA, and by TYLVD in all continents, apart from Antarctica. Because it devastates production of a valuable export crop, a sixth example caused by CSSD threatens livelihoods in West Africa. In addition, brief accounts are provided of other major virus disease epidemics arising from spillover scenarios involving new encounters in different parts of the world, namely MSD, MRDD, RHBD and several regional tomato orthotospovirus and begomovirus disease epidemics. For each pandemic or epidemic, the major factors driving its emergence initially, and its sudden increase in importance and geographical distribution subsequently, are explained. All these examples illustrate how spillover event initiated virus diseases epidemics that threaten food security in developing countries, vary greatly. This variation depends upon the characteristics of the causal virus(es), the crop affected, and the diversity of virus transmission modes, disease cycles, epidemiology, agro-ecological production systems and climatic factors involved. Tackling them successfully requires collaboration between policy makers, funding agencies, researchers and extension personnel on intercontinental scales. This scale of activity is needed to obtain a full understanding of each virus–vector–crop pathosystem and its epidemic drivers, and develop effective control measure approaches and extension strategies. Due to the urgent need to feed the growing global population, and address the increasing difficulties in controlling virus diseases of both staple and other food crops effectively as climate change progresses, the importance of the task ahead should not be underestimated.

Author Contributions: R.A.C.J. initiated and wrote this review. The author has read and agreed to the published version of the manuscript.

Funding: This review was written using resources provided by the Institute of Agriculture, University of Western Australia.

Acknowledgments: I thank Maruthi Gowda, Natural Resources Institute, University of Greenwich, UK for supplying the Figure 1 images of CBSD; and Naidu Rayapati, Irrigated Agriculture Research and Extension Center, Washington State University, Prosser, USA, and Lava Kumar, Germplasm Health Unit, Institute of Tropical Agriculture, Ibadan, Nigeria for supplying the Figure 2 images of GRD and CSSD, respectively. I also thank the The Annual Review of Virology for permission to reuse Figure 1D (CMD in cassava) from Jones and Naidu (2019), Annu. Rev. Virol. 6, 387–409, as Figure 1A, and the Annals of Applied Biology for permission to reuse Figure 1E (TYLVD in tomato) from Jones (2014), Ann. Appl. Biol. 164, 320–347, as Figure 1E. I also thank Michael Goodin, Adrian Gibbs and Martin Barbetti for their encouragement to write this review, and the Institute of Agriculture, University of Western Australia for use of its resources.

Conflicts of Interest: The author declares he has no conflict of interest.

References

1. Thresh, J.M. The origins and epidemiology of some important plant virus diseases. *Appl. Biol.* **1980**, *5*, 1–65.
2. Thresh, J.M. Crop viruses and virus diseases: A global perspective. In *Virus Diseases and Crop Biosecurity*; Cooper, J.I., Kuehne, T., Polischuk, V.P., Eds.; Springer: Dordrecht, The Netherlands, 2006; pp. 9–32.
3. Anderson, P.K.; Cunningham, A.A.; Patel, N.G.; Morales, F.J.; Epstein, P.R.; Daszak, P. Emerging infectious diseases of plants: Pathogen pollution, climate change and agrotechnology drivers. *Trends Ecol. Evol.* **2004**, *19*, 535–544. [CrossRef]
4. Jones, R.A.C. Plant virus emergence and evolution: Origins, new encounter scenarios, factors driving emergence, effects of changing world conditions, and prospects for control. *Virus Res.* **2009**, *141*, 113–130. [CrossRef] [PubMed]

5. Jones, R.A.C. Plant virus ecology and epidemiology: Historical perspectives, recent progress and future prospects. *Ann. Appl. Biol.* **2014**, *164*, 320–347. [CrossRef]

6. Hull, R. *Mathews' Plant Virology*, 5th ed.; Academic Press: London, UK, 2014.

7. Jones, R.A.C.; Naidu, R.A. Global dimensions of plant virus diseases: Current status and future perspectives. *Annu. Rev. Virol.* **2019**, *6*, 387–409. [CrossRef] [PubMed]

8. Bos, L. Wild plants in the ecology of virus diseases. In *Plant Diseases and Vectors: Ecology and Epidemiology*; Maramarosch, K., Harris, K.F., Eds.; Academic Press: New York, NY, USA, 1981; pp. 1–8.

9. Thresh, J.M. (Ed.) *Pests, Pathogens and Vegetation*; Pitman: London, UK, 1981.

10. Cooper, J.I.; Jones, R.A.C. Wild plants and viruses: Under-investigated ecosystems. *Adv. Virus Res.* **2006**, *67*, 1–47.

11. Alexander, H.M.; Mauck, K.E.; Whitfield, A.M.; Garrett, K.A.; Malmstrom, C.M. Plant-virus interactions and the agro-ecological interface. *Eur. J. Plant Pathol.* **2014**, *138*, 529–547. [CrossRef]

12. Jones, R.A.C.; Coutts, B.A. Spread of introduced viruses to new plants in natural ecosystems and the threat this poses to plant biodiversity. *Mol. Plant Pathol.* **2015**, *16*, 541–545. [CrossRef] [PubMed]

13. Malmstrom, C.M.; Alexander, H.M. Effects of crop viruses on wild plants. *Curr. Opin. Virol.* **2016**, *19*, 30–36. [CrossRef]

14. Bos, L. Crop losses caused by viruses. *Crop Prot.* **1982**, *1*, 263–282. [CrossRef]

15. Jones, R.A.C. Using epidemiological information to develop effective integrated virus disease management strategies. *Virus Res.* **2004**, *100*, 5–30. [CrossRef] [PubMed]

16. Jones, R.A.C. Control of plant virus diseases. *Adv. Virus Res.* **2006**, *67*, 205–244. [PubMed]

17. Sastry, S.K.; Zitter, T.A. Management of virus and viroid diseases of crops in the tropics. In *Plant Virus and Viroid Diseases in the Tropics, Volume 2, Epidemiology and Management*; Sastry, S.K., Zitter, T.A., Eds.; Springer: Dordrecht, The Netherlands, 2014; pp. 149–480.

18. Jeger, M.J.; Thresh, J.M. Modelling reinfection of replanted cocoa by swollen shoot virus in pandemically diseased areas. *J. Appl. Ecol.* **1993**, *30*, 187–196. [CrossRef]

19. Thresh, J.M. Control of tropical plant virus diseases. *Adv. Virus Res.* **2006**, *67*, 245–295.

20. Thresh, J.M. Plant virus epidemiology: The concept of host genetic vulnerability. *Adv. Virus Res.* **2006**, *67*, 89–125.

21. Brunt, A.; Crabtree, K.; Gibbs, A.J. *Viruses of Tropical Plants: Descriptions and Lists from the VIDE Database*; CAB International: Wallingford, UK, 1990.

22. Bos, L. New plant virus problems in developing countries: A corollary of agricultural modernization. *Adv. Virus Res.* **1992**, *41*, 349–407.

23. Otim-Nape, G.W.; Thresh, J.M. The current pandemic of cassava mosaic virus disease in Uganda. In *The Epidemiology of Plant Diseases*; Jones, G., Ed.; Kluwer: Dordrecht, The Netherlands, 1998; pp. 423–443.

24. Mahuku, G.; Lockhart, B.E.; Wanjala, B.; Jones, M.W.; Kimunye, J.N.; Stewart, L.R.; Cassone, B.J.; Sevgan, S.; Nyasani, J.O.; Kusia, E.; et al. Maize lethal necrosis (MLN), an emerging threat to maize-based food security in sub-Saharan Africa. *Phytopathology* **2015**, *105*, 956–965. [CrossRef]

25. Oliver, J.E.; Whitfield, A.E. The genus *Tospovirus*: Emerging bunyaviruses that threaten food security. *Annu. Rev. Virol.* **2016**, *3*, 101–124. [CrossRef]

26. Catherall, P.L. Effects of barley yellow dwarf and ryegrass mosaic viruses alone and in combination on the productivity of perennial and Italian ryegrass. *Plant Pathol.* **1987**, *36*, 73–78. [CrossRef]

27. McLaughlin, M.R.; Pederson, G.A.; Evans, R.R.; Ivy, R.L. Virus diseases and stand decline in a white clover pasture. *Plant Dis.* **1992**, *76*, 158–162. [CrossRef]

28. Jones, R.A.C.; Nicholas, D.A. Impact of an insidious virus disease in the legume component on the species balance within self-regenerating annual pasture. *J. Agric. Sci. Camb.* **1998**, *131*, 155–170. [CrossRef]

29. Eagling, D.R.; Villalta, O.; Sward, R.J. Host range, symptoms and effects on pasture production of a Victorian isolate of ryegrass mosaic potyvirus. *Aust. J. Agric. Res.* **1992**, *43*, 1243–1251. [CrossRef]

30. Coutts, B.A.; Jones, R.A.C. Temporal dynamics of spread of four viruses within mixed species perennial pastures. *Ann. Appl. Biol.* **2002**, *140*, 37–52. [CrossRef]

31. Jones, R.A.C. Virus diseases of annual pasture legumes: Incidences, losses, epidemiology and management. *Crop Pasture Sci.* **2012**, *63*, 399–418. [CrossRef]

32. Jones, R.A.C. Virus diseases of perennial pasture legumes: Incidences, losses, epidemiology and management. *Crop Pasture Sci.* **2013**, *64*, 199–215. [CrossRef]

33. Jones, R.A.C. Virus diseases of pasture grasses in Australia: Incidences, losses, epidemiology and management. *Crop Pasture Sci.* **2013**, *64*, 216–233. [CrossRef]

34. Malmstrom, C.M.; Hughes, C.C.; Newton, L.A.; Stoner, C.L. Virus infection in remnant native bunchgrasses from invaded California grasslands. *New Phytol.* **2005**, *168*, 217–230. [CrossRef]

35. Malmstrom, C.M.; McCullough, A.J.; Johnson, H.A.; Newton, L.A.; Borer, E.T. Invasive annual grasses indirectly increase virus incidence in California native perennial bunchgrasses. *Oecologia* **2005**, *145*, 153–164. [CrossRef]

36. Malmstrom, C.M.; Bigelow, P.; Trębicki, P.; Busch, A.K.; Friel, C.; Cole, E.; Abdel-Azim, H.; Phillippo, C.; Alexander, H.M. Crop-associated virus reduces the rooting depth of non-crop perennial native grass more than non-crop-associated virus with known viral suppressor of RNA silencing (VSR). *Virus Res.* **2017**, *241*, 172–184. [CrossRef]

37. Vincent, S.J.; Coutts, B.A.; Jones, R.A.C. Effects of introduced and indigenous viruses on native plants: Exploring their disease causing potential at the agro-ecological interface. *PLoS ONE* **2014**, *9*, e91224. [CrossRef]

38. Thresh, J.M. Control of plant virus diseases in sub-Saharan Africa: The possibility and feasibility of an integrated approach. *Afr. Crop Sc. J.* **2003**, *11*, 199–223. [CrossRef]

39. Loebenstein, G.; Thottappilly, G. (Eds.) *Virus and Virus-like Diseases of Major Crops in Developing Countries*; Springer: Dordrecht, The Netherlands, 2013.

40. Thresh, J.M. Cropping practices and virus spread. *Annu. Rev. Phytopathol.* **1982**, *20*, 193–218. [CrossRef]

41. Jones, R.A.C.; Barbetti, M.J. Influence of climate change on plant disease infections and epidemics caused by viruses and bacteria. *Plant Sci. Rev.* **2012**, *22*, 1–31. [CrossRef]

42. Roossinck, M.J.; García-Arenal, F. Ecosystem simplification, biodiversity loss and plant virus emergence. *Curr. Opin. Virol.* **2015**, *10*, 56–62. [CrossRef]

43. Webster, C.G.; Coutts, B.A.; Jones, R.A.C.; Jones, M.G.K.; Wylie, S.J. Virus impact at the interface of an ancient ecosystem and a recent agroecosystem: Studies on three legume-infecting potyviruses in the South West Australian Floristic Region. *Plant Pathol.* **2007**, *56*, 729–742. [CrossRef]

44. Faillace, C.A.; Lorusso, N.S.; Duffy, S. Overlooking the smallest matter: Viruses impact biological invasions. *Ecol. Lett.* **2017**, *20*, 524–538. [CrossRef]

45. Irwin, M.E.; Fereres, A. John Michael Thresh, founding father of plant virus epidemiology: A tribute. *Virus Res.* **2017**, *241*, 3–9. [CrossRef]

46. Thresh, J.M. An ecological approach to the epidemiology of plant virus diseases. In *Comparative Epidemiology: A Tool for Better Disease Management*; Palti, J., Kranz, J., Eds.; Centre for Agricultural Publishing and Documentation: Wageningen, The Netherlands, 1980; pp. 57–70.

47. Thresh, J.M. Plant virus epidemiology and control. Current trends and future prospects. In *Plant Virus Epidemiology—The Spread and Control of Insect-Borne Pathogens*; Plumb, R.T., Thresh, J.M., Eds.; Blackwell: Oxford, UK, 1983; pp. 349–360.

48. Thresh, J.M. Insect-borne viruses of rice and the Green Revolution. *Int. J. Pest Manag.* **1989**, *35*, 264–272. [CrossRef]

49. Thresh, J.M. Plant virus epidemiology: The battle of the genes. In *Recognition and Response in Plant-Virus Interactions*; [NATO ASI Series H: Cell Biology]; Springer: Heidelberg, Germany, 1990; Volume 41, pp. 93–121.

50. Thresh, J.M. The ecology of tropical plant viruses. *Plant Pathol.* **1991**, *40*, 324–339. [CrossRef]

51. Dombrovsky, A.; Tran-Nguyen, L.T.T.; Jones, R.A.C. *Cucumber Green Mottle Mosaic Virus*: Rapidly increasing global distribution, etiology, epidemiology, and management. *Annu. Rev. Phytopathol.* **2017**, *55*, 231–256. [CrossRef]

52. Legg, J.P.; Jeremiah, S.C.; Obiero, H.M.; Maruthi, M.N.; Ndyetabula, I.; Okao-Okuja, G.; Bouwmeester, H.; Bigirimana, S.; Tata-Hangy, W.; Gashaka, G.; et al. Comparing the regional epidemiology of the cassava mosaic and cassava brown streak virus pandemics in Africa. *Virus Res.* **2011**, *159*, 161–170. [CrossRef] [PubMed]

53. Moriones, E.; Praveen, S.; Chakraborty, S. *Tomato leaf curl New Delhi virus*: An emerging virus complex threatening vegetable and fibre crops. *Viruses* **2017**, *9*, 264. [CrossRef] [PubMed]

54. Rey, C.; Vanderschuren, H. Cassava mosaic and brown streak diseases: Current perspectives and beyond. *Annu. Rev. Virol.* **2017**, *4*, 429–452. [CrossRef] [PubMed]

55. Fargette, D.; Konate, G.; Fauquet, C.; Muller, E.; Peterscmitt, M.; Thresh, J.M. Molecular ecology and emergence of tropical plant viruses. *Ann. Rev. Phytopathol.* **2006**, *44*, 235–260. [CrossRef] [PubMed]

56. Redinbaugh, M.G.; Stewart, L.R. Maize lethal necrosis: An emerging, synergistic viral disease. *Annu. Rev. Virol.* **2018**, *5*, 301–322. [CrossRef]

57. Tomlinson, K.R.; Bailey, A.M.; Alicai, T.; Seal, S.; Foster, G.D. Cassava brown streak disease: Historical timeline, current knowledge and future prospects. *Mol. Plant Pathol.* **2018**, *19*, 1282–1294. [CrossRef]

58. Fiallo-Olivé, E.; Navas-Castillo, J. *Tomato chlorosis virus*, an emergent plant virus still expanding its geographical and host ranges. *Mol. Plant Pathol.* **2019**, *20*, 1307–1320. [CrossRef]

59. Klinkowski, M. Catastrophic plant diseases. *Ann. Rev. Phytopathol.* **1970**, *8*, 37–60. [CrossRef]

60. Cooper, J.I.; Jones, A.T. Responses of plants to viruses: Proposals for the use of terms. *Phytopathology* **1983**, *73*, 127–128. [CrossRef]

61. Harlan, J.R. Agricultural origins: Centres and non-centres. *Science* **1971**, *174*, 468–474. [CrossRef]

62. Harlan, J.R. Ecological settings for the emergence of agriculture. In *Pests, Pathogens and Vegetation*; Thresh, J.M., Ed.; Pitman Press: London, UK, 1981; pp. 3–22.

63. Nunn, N.; Qian, N. The Columbian Exchange: A history of disease, food, and ideas. *J. Econ. Perspect.* **2010**, *24*, 163–188. [CrossRef]

64. Elena, S.F.; Fraile, A.; Garcia-Arenal, F. Evolution and emergence of plant viruses. *Adv. Virus Res.* **2014**, *88*, 161–191. [PubMed]

65. Rojas, M.R.; Gilbertson, R.L. Emerging plant viruses: A diversity of mechanisms and opportunities. In *Plant Virus Evolution*; Roossinck, M.A., Ed.; Springer: Berlin/Heidelberg, Germany, 2008; pp. 27–51.

66. Pagán, I.; González-Jara, P.; Moreno-Letelier, A.; Rodelo-Urrego, M.; Fraile, A.; Piñero, D.; García-Arenal, F. Effect of biodiversity changes in disease risk: Exploring disease emergence in a plant-virus system. *PLoS Pathog.* **2012**, *8*, e1002796. [CrossRef] [PubMed]

67. Fereres, A. Insect vectors as drivers of plant virus emergence. *Curr. Opin. Virol.* **2015**, *10*, 42–46. [CrossRef] [PubMed]

68. Gilbertson, R.L.; Batuman, O.; Webster, C.G.; Adkins, S. Role of the insect supervectors *Bemisia tabaci* and *Frankliniella occidentalis* in the emergence and global spread of plant viruses. *Annu. Rev. Virol.* **2015**, *2*, 67–93. [CrossRef]

69. Jones, R.A.C. Future scenarios for plant virus pathogens as climate change progresses. *Adv. Virus Res.* **2016**, *95*, 87–147.

70. Jones, R.A.C. Plant and insect viruses in managed and natural environments: Novel and neglected transmission pathways. *Adv. Virus Res.* **2018**, *101*, 149–187.

71. García-Arenal, F.; Zerbini, F.M. Life on the edge: Geminiviruses at the interface between crops and wild plant hosts. *Annu. Rev. Virol.* **2019**, *6*, 411–433. [CrossRef]

72. McLeish, M.J.; Fraile, A.; García-Arenal, F. Evolution of plant–virus interactions: Host range and virus emergence. *Curr. Opin. Virol.* **2019**, *34*, 50–55. [CrossRef]

73. Lefeuvre, P.; Martin, D.P.; Elena, S.F.; Shepherd, D.N.; Roumagnac, P.; Varsani, A. Evolution and ecology of plant viruses. *Nat. Rev. Microbiol.* **2019**, *17*, 632–644. [CrossRef]

74. Sweeney, M.; McCouch, S. The complex history of the domestication of rice. *Ann. Bot.* **2007**, *100*, 951–957. [CrossRef] [PubMed]

75. Pinel-Galzi, A.; Traoré, O.; Séré, Y.; Hébrard, E.; Fargette, D. The biogeography of viral emergence: *Rice yellow mottle virus* as a case study. *Curr. Opin. Virol.* **2015**, *10*, 7–13. [CrossRef] [PubMed]

76. Seck, P.A.; Diagne, A.; Mohanty, S.; Wopereis, M.C. Crops that feed the world 7: Rice. *Food Secur.* **2012**, *4*, 7–24. [CrossRef]

77. FAOSTAT. *Food and Agricultural Organization Statistical Databases*; United Nations: Rome, Italy, 2020.

78. Hibino, H. Biology and epidemiology of rice viruses. *Ann. Rev. Phytopathol.* **1996**, *34*, 249–274. [CrossRef] [PubMed]

79. Traoré, O.; Fargette, D. Rice yellow mottle virus. In *Vector-Mediated Transmission of Plant Pathogens, Chapter 29*; APS Publications: St. Paul, MN, USA, 2016; pp. 425–429.

80. CABI. *CABI Compendium Datasheets*; CABI: Wallingford, UK, 2020.

81. Rakotomalala, M.; Vrancken, B.; Pinel-Galzi, A.; Ramavovololona, P.; Hébrard, E.; Randrianangaly, J.S.; Dellicour, S.; Lemey, P.; Fargette, D. Comparing patterns and scales of plant virus phylogeography: *Rice yellow mottle virus* in Madagascar and in continental Africa. *Virus Evol.* **2019**, *5*, vez023. [CrossRef]

82. Legg, J.P.; Kumar, P.L.; Makeshkumar, T.; Tripathi, L.; Ferguson, M.; Kanju, E.; Ntawuruhunga, P.; Cuellar, W. Cassava virus diseases: Biology, epidemiology, and management. *Adv. Virus Res.* **2015**, *91*, 85–142.

83. Legg, J.; Somado, E.A.; Barker, I.; Beach, L.; Ceballos, H.; Cuellar, W.; Elkhoury, W.; Gerling, D.; Helsen, J.; Hershey, C.; et al. A global alliance declaring war on cassava viruses in Africa. *Food Secur.* **2014**, *6*, 231–248. [CrossRef]

84. Siriwan, W.; Jimenez, J.; Hemniam, N.; Saokham, K.; Lopez-Alvarez, D.; Leiva, A.M.; Martinez, A.; Mwanzia, L.; Becerra, L.A.; Cuellar, W.J. Surveillance and Diagnostics of the emergent Sri Lankan cassava mosaic virus (Fam. Geminiviridae) in Southeast Asia. *Virus Res.* **2020**, *285*, 197959. [CrossRef]

85. Dzahini-Obiatey, H.; Domfeh, O.; Amoah, F.M. Over seventy years of a viral disease of cocoa in Ghana: From researchers perspective. *Afr. J. Agric. Res.* **2010**, *5*, 476–485.

86. Ameyaw, G.A. Management of the cacao swollen shoot virus (CSSV). Menace in Ghana: The past, present and the future. In *Plant Diseases-Current Threats and Management Trends*; Topolovec-Pintarić, S., Ed.; IntechOpen: London, UK, 2019; pp. 1–13. [CrossRef]

87. Naidu, R.A.; Bottenberg, H.; Subrahmanyam, P.; Kimmins, F.M.; Robinson, D.J.; Thresh, J.M. Epidemiology of groundnut rosette virus disease: Current status and future research needs. *Ann. Appl. Biol.* **1998**, *132*, 525–548. [CrossRef]

88. Naidu, R.A.; Kimmins, F.M.; Deom, C.M.; Subrahmanyam, P.; Chiyembekeza, A.J.; van der Merwe, P.J.A. Groundnut rosette: A virus disease affecting groundnut production in sub-saharan Africa. *Plant Dis.* **1999**, *83*, 700–709. [CrossRef] [PubMed]

89. Taliansky, M.E.; Robinson, D.J.; Murant, A.F. Groundnut rosette disease virus complex: Biology and molecular biology. *Adv. Virus Res.* **2000**, *56*, 357–400.

90. Mabele, A.S.; Were, H.K.; Ndong'a, M.F.; Mukoye, B. Distribution, molecular detection and host range of groundnut rosette assistor virus in western Kenya. *J. Plant Sci.* **2019**, *7*, 100–105.

91. Moriones, E.; Navas-Castillo, J. *Tomato yellow leaf curl virus*, an emerging virus complex causing epidemics worldwide. *Virus Res.* **2000**, *71*, 123–134. [CrossRef]

92. Mabvakure, B.; Martin, D.P.; Kraberger, S.; Cloete, L.; van Brunschot, S.; Geering, A.D.; Thomas, J.E.; Bananej, K.; Lett, J.M.; Lefeuvre, P.; et al. Ongoing geographical spread of *Tomato yellow leaf curl virus*. *Virology* **2016**, *498*, 257–264. [CrossRef]

93. Prasad, A.; Sharma, N.; Hari-Gowthem, G.; Muthamilarasan, M.; Prasad, M. *Tomato yellow leaf curl virus*: Impact, challenges, and management. *Trends Plant Sci.* **2020**, *25*, 897–911. [CrossRef]

94. Luisoni, E.; Conti, M. *Digitaria sanguinalis* (L.) Scop., a new natural host of maize rough dwarf virus. *Phytopathol. Mediterr.* **1970**, *9*, 102–105.

95. Harpaz, I. (Ed.) Maize rough dwarf. In *A Planthopper Virus Disease Affecting Maize, Rice, Small Grains and Grasses*; Israel Universities Press: Jerusalem, Israel, 1972; p. 251.

96. Achon, M.A.; Serrano, L.; Clemente-Orta, G.; Barcelo, A. The virome of maize rough dwarf disease: Molecular genome diversification, phylogeny and selection. *Ann. Appl. Biol.* **2020**, *176*, 192–202. [CrossRef]

97. Bosque-Pérez, N.A. Eight decades of maize streak virus research. *Virus Res.* **2000**, *71*, 107–121. [CrossRef]

98. Shepherd, D.N.; Martin, D.P.; Van der Walt, E.; Dent, K.; Varsani, A.; Rybicki, E.P. Maize streak virus: An old and complex 'emerging' pathogen. *Mol. Plant Pathol.* **2010**, *11*, 1–12. [CrossRef]

99. Morales, F.J.; Jennings, P.R. Rice hoja blanca: A complex plant-virus-vector pathosystem. *Plant Sci. Rev.* **2010**, *5*, 1–16. [CrossRef]

100. Sarra, S.; Peters, D. Rice yellow mottle virus is transmitted by cows, donkeys, and grass rats in irrigated rice crops. *Plant Dis.* **2003**, *87*, 804–808. [CrossRef] [PubMed]

101. Parmar, A.; Sturm, B.; Hensel, O. Crops that feed the world: Production and improvement of cassava for food, feed, and industrial uses. *Food Secur.* **2017**, *9*, 907–927. [CrossRef]

102. Zhou, X.; Liu, Y.; Calvert, L.; Munoz, C.; Otim-Nape, G.W.; Robinson, D.J.; Harrison, B.D. Evidence that DNA-A of a geminivirus associated with severe cassava mosaic disease in Uganda has arisen by interspecific recombination. *J. Gen. Virol.* **1997**, *78*, 2101–2111. [CrossRef]

103. Legg, J.; Fauquet, C. Cassava mosaic geminiviruses in Africa. *Plant Mol. Biol.* **2004**, *56*, 585–599. [CrossRef]

104. Patil, B.; Fauquet, C. Cassava mosaic geminiviruses: Actual knowledge and perspectives. *Mol. Plant Pathol.* **2009**, *10*, 685–701. [CrossRef]

105. Thresh, J.M.; Fargette, D.; Otim-Nape, G.W. Effects of African cassava mosaic geminivirus on the yield of cassava. *Trop. Sci. Lond.* **1994**, *34*, 26.

106. Legg, J.P.; Sseruwagi, P.; Boniface, S.; Okao-Okuja, G.; Shirima, R.; Bigirimana, S.; Gashaka, G.; Herrmann, H.W.; Jeremiah, S.; Obiero, H.; et al. Spatio-temporal patterns of genetic change amongst populations of cassava *Bemisia tabaci* whiteflies driving virus pandemics in East and Central Africa. *Virus Res.* **2014**, *186*, 61–75. [CrossRef]

107. Sserubombwe, W.S.; Briddon, R.W.; Baguma, Y.K.; Ssemakula, G.N.; Bull, S.E.; Bua, A.; Alicai, T.; Omongo, C.; Otim-Nape, G.W.; Stanley, J. Diversity of begomoviruses associated with mosaic disease of cultivated cassava (*Manihot esculenta* Crantz) and its wild relative (*Manihot glaziovii* Müll. Arg.) in Uganda. *J. Gen. Virol.* **2008**, *89*, 1759–1769. [CrossRef]

108. Ndunguru, J.; De León, L.; Doyle, C.D.; Sseruwagi, P.; Plata, G.; Legg, J.P.; Thompson, G.; Tohme, J.; Aveling, T.; Ascencio-Ibáñez, J.T.; et al. Two novel DNAs that enhance symptoms and overcome CMD2 resistance to cassava mosaic disease. *J. Virol.* **2016**, *90*, 4160–4173. [CrossRef]

109. Shirima, R.R.; Maeda, D.G.; Kanju, E.E.; Tumwegamire, S.; Ceasar, G.; Mushi, E.; Sichalwe, C.; Mtunda, K.; Mkamilo, G.; Legg, J.P. Assessing the degeneration of cassava under high-virus inoculum conditions in coastal Tanzania. *Plant Dis.* **2019**, *103*, 2652–2664. [CrossRef] [PubMed]

110. Hillocks, R.J.; Raya, M.D.; Mtunda, K.; Kiozia, H. Effects of brown streak virus disease on yield and quality of cassava in Tanzania. *J. Phytopathol.* **2001**, *149*, 389–394. [CrossRef]

111. Maruthi, M.N.; Hillocks, R.J.; Mtunda, K.; Raya, M.D.; Muhanna, M.; Kiozia, H.; Rekha, A.R.; Colvin, J.; Thresh, J.M. Transmission of *Cassava brown streak virus* by *Bemisia tabaci* (*Gennadius*). *J. Phytopathol.* **2005**, *153*, 307–312. [CrossRef]

112. Maruthi, M.N.; Jeremiah, S.C.; Mohammed, I.U.; Legg, J.P. The role of the whitefly, *Bemisia tabaci* (*Gennadius*), and farmer practices in the spread of cassava brown streak ipomoviruses. *J. Phytopathol.* **2017**, *165*, 707–717. [CrossRef]

113. Monger, W.A.; Alicai, T.; Ndunguru, J.; Kinyua, Z.M.; Potts, M.; Reeder, R.H.; Miano, D.W.; Adams, I.P.; Boonham, N.; Glover, R.H.; et al. The complete genome sequence of the Tanzanian strain of *Cassava brown streak virus* and comparison with the Ugandan strain sequence. *Arch. Virol.* **2010**, *155*, 429–433. [CrossRef] [PubMed]

114. Alicai, T.; Ndunguru, J.; Sseruwagi, P.; Tairo, F.; Okao-Okuja, G.; Nanvubya, R.; Kiiza, L.; Kubatko, L.; Kehoe, M.A.; Boykin, L.M. Characterization by next generation sequencing reveals the molecular mechanisms driving the faster evolutionary rate of *Cassava brown streak virus* compared with Ugandan cassava brown streak virus. *Sci. Rep.* **2016**, *6*, 36164. [CrossRef] [PubMed]

115. Amisse, J.J.; Ndunguru, J.; Tairo, F.; Boykin, L.M.; Kehoe, M.A.; Cossa, N.; Ateka, E.; Rey, C.; Sseruwagi, P. First report of cassava brown streak viruses on wild plant species in Mozambique. *Physiol. Mol. Plant Pathol.* **2019**, *105*, 88–95. [CrossRef]

116. Jeremiah, S.C.; Ndyetabula, I.L.; Mkamilo, G.S.; Haji, S.; Muhanna, M.M.; Chuwa, C.; Kasele, S.; Bouwmeester, H.; Ijumba, J.N.; Legg, J.P. The dynamics and environmental influence on interactions between Cassava brown streak disease and the whitefly, *Bemisia tabaci*. *Phytopathology* **2015**, *105*, 646–655. [CrossRef]

117. Jenkins, J.A. The origin of the cultivated tomato. *Econ. Bot.* **1948**, *2*, 379–392. [CrossRef]

118. Hanssen, I.M.; Lapidot, M.; Thomma, B.P.H.J. Emerging viral diseases of tomato crops. *Mol. Plant-Microbe Interact.* **2010**, *23*, 539–548. [CrossRef]

119. Rojas, M.R.; Macedo, M.A.; Maliano, M.R.; Soto-Aguilar, M.; Souza, J.O.; Briddon, R.W.; Kenyon, L.; Rivera-Bustamante, R.F.; Zerbini, F.M.; Adkins, S.; et al. World management of geminiviruses. *Annu. Rev. Phytopathol.* **2018**, *56*, 637–677. [CrossRef] [PubMed]

120. Kil, E.J.; Kim, S.; Lee, Y.J.; Byun, H.S.; Park, J.; Seo, H.; Kim, C.S.; Shim, J.K.; Lee, J.H.; Kim, J.K.; et al. Tomato yellow leaf curl virus (TYLCV-IL): A seed-transmissible geminivirus in tomatoes. *Sci. Rep.* **2016**, *6*, 19013. [CrossRef] [PubMed]

121. Lefeuvre, P.; Martin, D.P.; Harkins, G.; Lemey, P.; Gray, A.J.; Meredith, S.; Lakay, F.; Monjane, A.; Lett, J.M.; Varsani, A.; et al. The spread of tomato yellow leaf curl virus from the Middle East to the world. *PLoS Pathog.* **2010**, *6*, e1001164. [CrossRef] [PubMed]

122. Polston, J.E.; Anderson, P.K. The emergence of whitefly-transmitted geminiviruses in tomato in the Western Hemisphere. *Plant Dis.* **1997**, *81*, 1358–1369. [CrossRef]

123. Monci, F.; Sánchez-Campos, S.; Navas-Castillo, J.; Moriones, E. A natural recombinant between the geminiviruses *Tomato yellow leaf curl Sardinia virus* and *Tomato yellow leaf curl virus* exhibits a novel pathogenic phenotype and is becoming prevalent in Spanish populations. *Virology* **2002**, *303*, 317–326. [CrossRef]

124. Díaz-Pendón, J.A.; Cañizares, M.C.; Moriones, E.; Bejarano, E.R.; Czosnek, H.; Navas-Castillo, J. Tomato yellow leaf curl viruses: Ménage à trois between the virus complex, the plant and the whitefly vector. *Mol. Plant Pathol.* **2010**, *11*, 441–450. [CrossRef]

125. Navas-Castillo, J.; Fiallo-Olivé, E.; Sánchez-Campos, S. Emerging virus diseases transmitted by whiteflies. *Annu. Rev. Phytopathol.* **2011**, *49*, 219–248. [CrossRef]

126. Kenyon, L.; Tsai, W.S.; Shih, S.L.; Lee, L.M. Emergence and diversity of begomoviruses infecting solanaceous crops in East and Southeast Asia. *Virus Res.* **2014**, *186*, 104–113. [CrossRef]

127. Sánchez-Campos, S.; Navas-Castillo, J.; Camero, R.; Soria, C.; Díaz, J.A.; Moriones, E. Displacement of tomato yellow leaf curl virus (TYLCV)-Sr by TYLCV-Is in tomato epidemics in Spain. *Phytopathology* **1999**, *89*, 1038–1043. [CrossRef]

128. Foyer, C.H.; Lam, H.M.; Nguyen, H.T.; Siddique, K.H.; Varshney, R.K.; Colmer, T.D.; Cowling, W.; Bramley, H.; Mori, T.A.; Hodgson, J.M.; et al. Neglecting legumes has compromised human health and sustainable food production. *Nat. Plants* **2016**, *22*, 16112. [CrossRef]

129. Bos, L. *Research on Viruses of Legume Crops and the International Working Group on Legume Viruses: Historical Facts and Personal Reminiscences*; ICARDA Publication: Aleppo, Syria, 1996.

130. Makkouk, K.M.; Kumari, S.G.; van Leur, J.A.G.; Jones, R.A.C. Control of plant virus diseases in cool-season grain legume crops. *Adv. Virus Res.* **2014**, *90*, 207–253. [PubMed]

131. Makkouk, K.M. Plant pathogens which threaten food security: Viruses of chickpea and other cool season legumes in West Asia and North Africa. *Food Secur.* **2020**, *12*, 495–502. [CrossRef]

132. Rachie, K.O.; Roberts, L.M. Grain Legumes of the Lowland Tropics. *Adv. Agron.* **1974**, *26*, 1–132.

133. Hema, M.; Sreenivasulu, P.; Patil, B.L.; Kumar, P.L.; Reddy, D.V. Tropical food legumes: Virus diseases of economic importance and their control. *Adv. Virus Res.* **2014**, *90*, 431–505. [PubMed]

134. Zhuang, W.; Chen, H.; Yang, M.; Wang, J.; Pandey, M.K.; Zhang, C.; Chang, W.C.; Zhang, L.; Zhang, X.; Tang, R.; et al. The genome of cultivated peanut provides insight into legume karyotypes, polyploid evolution and crop domestication. *Nat. Genet.* **2019**, *51*, 865–876. [CrossRef]

135. Crowdy, S.H.; Posnette, A.F. Virus diseases of cacao in West Africa. II Cross-immunity experiments with viruses 1A, 1B, 1C. *Ann. Appl. Biol.* **1947**, *34*, 403–411. [CrossRef]

136. Tinsley, T.W. The ecology of cacao viruses. I. The role of wild hosts in the incidence of swollen shoot virus in West Africa. *J. Appl. Ecol.* **1971**, *8*, 491–495. [CrossRef]

137. Dongo, L.N.; Orisajo, S.B. Status of cocoa swollen shoot virus disease in Nigeria. *Afr. J. Biotechnol.* **2007**, *6*, 2054–2061.

138. Shiferaw, B.; Prasanna, B.M.; Hellin, J.; Bänziger, M. Crops that feed the world 6. Past successes and future challenges to the role played by maize in global food security. *Food Secur.* **2011**, *3*, 307–327. [CrossRef]

139. McCann, J. Maize and grace: History, corn, and Africa's new landscapes, 1500–1999. *Comp. Stud. Soc. Hist.* **2001**, *43*, 246–272. [CrossRef]

140. Bateson, M.F.; Lines, R.E.; Revill, P.; Chaleeprom, W.; Ha, C.V.; Gibbs, A.J.; Dale, J.L. On the evolution and molecular epidemiology of the potyvirus. *J. Gen. Virol.* **2002**, *83*, 2575–2585. [CrossRef]

141. Pappu, H.R.; Jones, R.A.C.; Jain, R.K. Global status of tospovirus epidemics in diverse cropping systems: Successes achieved and challenges ahead. *Virus Res.* **2009**, *141*, 219–236. [CrossRef] [PubMed]

142. Mandal, B.; Jain, R.K.; Krishnareddy, M.; Krishna Kumar, N.K.; Ravi, K.S.; Pappu, H.R. Emerging problems of tospoviruses (*Bunyaviridae*) and their management in the Indian subcontinent. *Plant Dis.* **2012**, *96*, 468–479. [CrossRef] [PubMed]

143. Rai, A.K.; Sadashiva, A.T.; Basavaraj, Y.B.; Venugopalan, R.; Rao, E.S.; Nandeesha, P. Genetic analysis of bud necrosis disease caused by groundnut bud necrosis virus (GBNV) in tomato (*Solanum lycopersicum* L.). *Euphytica* **2020**, *216*, 1–10. [CrossRef]

144. Karmakar, A.; Saha, D.; Saha, S. Mini review on *Begomoviruses* infecting cultivated vegetable crops in India. In *Organisms and Environment*; Sarkar, A.N., Ed.; Education publishing: New Delhi, India, 2020; pp. 176–185.

145. Yazdani-Khameneh, S.; Golnaraghi, A. The status of *Begomoviruses* in Iran. In *Begomoviruses: Occurrence and Management in Asia and Africa*; Saxena, S., Tiwari, A., Eds.; Springer: Singapore, 2017; pp. 229–253.

146. Leke, W.N.; Mignouna, D.B.; Brown, J.K.; Kvarnheden, A. *Begomovirus* disease complex: Emerging threat to vegetable production systems of West and Central Africa. *Agric. Food Secur.* **2015**, *4*, 1–14. [CrossRef]

147. Morales, F.J.; Jones, P.G. The ecology and epidemiology of whitefly-transmitted viruses in Latin America. *Virus Res.* **2004**, *100*, 57–65. [CrossRef] [PubMed]

148. Morales, F.J. History and current distribution of begomoviruses in Latin America. *Adv. Virus Res.* **2006**, *67*, 127–162.

149. Rocha, C.S.; Castillo-Urquiza, G.P.; Lima, A.T.; Silva, F.N.; Xavier, C.A.; Hora-Júnior, B.T.; Beserra-Júnior, J.E.; Malta, A.W.; Martin, D.P.; Varsani, A.; et al. Brazilian begomovirus populations are highly recombinant, rapidly evolving, and segregated based on geographical location. *J. Virol.* **2013**, *87*, 5784–5799.

150. Inoue-Nagata, A.K.; Lima, M.F.; Gilbertson, R.L. A review of geminivirus diseases in vegetables and other crops in Brazil: Current status and approaches for management. *Hortic. Bras.* **2016**, *34*, 8–18. [CrossRef]

151. Barboza, N.M.; Hernández, E.; Inoue-Nagata, A.K.; Moriones, E.; Hilje, L. Achievements in the epidemiology of begomoviruses and their vector *Bemisia tabaci* in Costa Rica. *Rev. Biol. Trop.* **2019**, *67*, 419–453.

152. Morales, F.J.; Anderson, P.K. The emergence and dissemination of whitefly-transmitted geminiviruses in Latin America. *Arch. Virol.* **2001**, *146*, 415–441. [CrossRef] [PubMed]

153. Rodoni, B. The role of plant biosecurity in preventing and controlling emerging plant virus disease epidemics. *Virus Res.* **2009**, *141*, 150–157. [CrossRef] [PubMed]

Publisher's Note: MDPI stays neutral with regard to jurisdictional claims in published maps and institutional affiliations.

Article

Emergence and Full Genome Analysis of *Tomato Torrado Virus* in South Africa

Vaneson Moodley [1],*, Augustine Gubba [1] and Paramu L. Mafongoya [2]

[1]   Discipline of Plant Pathology, School of Agricultural, Earth and Environmental Sciences,
      University of KwaZulu-Natal, Scottsville, Pietermaritzburg 3209, South Africa; GubbaA@ukzn.ac.za
[2]   Rural Agronomy and Development, School of Agricultural, Earth and Environmental Sciences,
      University of KwaZulu-Natal, Scottsville, Pietermaritzburg 3209, South Africa; mafongoya@ukzn.ac.za
*   Correspondence: vaneson.moodley@gmail.com

Received: 1 September 2020; Accepted: 22 September 2020; Published: 15 October 2020

Abstract: Emerging pests and diseases are a major threat to food production worldwide. In a recent survey, *Tomato torrado virus* (ToTV) was identified on tomato crops in the Limpopo province of South Africa and a first report of the disease was published. In this follow-up study, the full genome sequence of a tomato-infecting isolate of ToTV from South Africa was elucidated. High-throughput sequencing was used to generate the full genome of ToTV infecting tomato crops in South Africa. The longest contig obtained for the RNA-1 and RNA-2 genome of ToTV was comprised of 7420 and 5381 nucleotides (nt), respectively. Blast analysis of the RNA-1 sequence of ToTV from South Africa (ToT-186) matched 99% to a Spanish and Polish isolate; the RNA-2 segment of ToTV from South Africa (ToT-186) matched 99% to ToTV isolates from Italy and Poland, respectively. The information presented in this study will go a long way towards better understanding the emergence and spread of ToTV and devising sustainable management of ToTV diseases.

Keywords: ToTV; emerging disease; prevalence; whole-genome sequencing; phylogeny

---

1. Introduction

Emerging pests and disease have destroyed agricultural crops all over the world. More than fifteen years ago, local farmers in the city of Murcia (located in the southeastern parts of Spain) observed severe necrosis on the leaves and fruit of tomato crops. These symptoms later became known as "torrado" disease which was coined by local Spanish farmers describing the "burnt-like" effect of the disease in tomato fields. A subsequent study by Verbeek et al. [1] provided a detailed analysis of a new picorna-like virus infecting tomatoes which they termed *"Tomato torrado virus"*. Although this newly discovered species was shown to display several characteristics similar to the Sequivirus, Sadwavirus, and Cheravirus genera in the Sequiviridae family, sequence characteristics distinguished *Tomato torrado virus* (ToTV) as a member of a new plant virus genus [1].

According to Sanfacon et al. [2]. ToTV is the type member of the genus Torradovirus in the family Secoviridae which is an amalgamation of the families Sequiviridae and Comoviridae, together with previously unassigned genera Cheravirus and Sadwavirus, in the order Picornavirales. In addition, members of the Secoviridae family have a small icosahedral particle morphology (25–30 nm) with a pseudo-T = 3 symmetry, and a mono/bipartite positive-strand RNA genome. The capsid of non-enveloped virions contain jelly-roll domains that are organized into three mature capsid proteins that are folded alike but vary in amino acid sequence and length [3]. These domains may have resulted from the triplication of a single domain and consecutive divergent evolution [4].

Torradoviruses have a bipartite genome composed of an RNA-1 (7.8 kb) and an RNA-2 (5.4 kb) segment. Each genomic segment has a VPg linked to its 5′ end and a 3′ poly (A) tract. RNA-1 and RNA-2 are translated into two polyproteins, which are then processed into functional proteins. RNA-1 encodes proteins (a type III helicase, 3C-like proteinase, and a type I RNA-dependent RNA-polymerase) that are necessary for replication while RNA-2 encodes three coat proteins and proteins involved in virus movement. It is further demonstrated that the genome of torradoviruses consists of an exclusive second open reading frame (ORF) upstream of RNA-2 which partially overlaps the large ORF and encodes a protein of unknown function that displays a great degree of sequence diversity with other torradoviruses [2]. The large 3′ NTR shares > 99% sequence identity between the RNA-1 and RNA-2 segments of a particular torradovirus species but varies substantially in terms of interspecific differences.

ToTV along with three other torradovirus species i.e., *Tomato marchitez virus* (ToMarV), *Tomato chocolàte virus* (ToChV), and *Tomato necrotic dwarf virus* (ToNDV) are presently the only known spherical viruses that are transmitted by three whitefly species, i.e., *Trialeurodes vaporariorum*, *Bemisia tabaci*, and *Trialeurodes abutilonea* in a semi-persistent manner [5,6]. Verbeek et al. [5] further demonstrate that the virus is retained in the stylet where it may remain for up to eight hours without loss of transmission efficiency. The presence of whitefly species has been identified to varying degrees throughout the world. Studies of emerging whitefly-transmitted viruses such as ToTV have been linked to abnormally high vector populations [7]. Jones et al. [8] indicate that *B. tabaci* and *T. vaporariorum* thrive in warmer temperatures with prolonged dry spells of up to four months (<80 mm rainfall/month), although *T. vaporariorum* is more tolerant of cold weather.

South Africa recently experienced a severe drought which was exacerbated by the effects of *El Niño*. Provinces situated in the north experienced extremely high temperatures and prolonged periods of dry weather that resulted in millions of dollars in crop losses. During a national survey, an unprecedented whitefly infestation was observed in major tomato growing areas throughout South Africa. Tomato growers in the Limpopo province (the largest producer of tomatoes in South Africa) noticed severe necrosis/burnt-like symptoms on the leaves, stems, and fruit that resembled heat burn or possibly osmotic stress that usually results from an excess of nitrogen salts (personal communication). Closer observation revealed that the "burnt like" symptoms were similar to those described by the tomato farmers in Murcia. Primary symptoms appeared as necrotic spots surrounded by chlorotic halos that began at the base of immature leaves, and advanced as stunted growth, vertical stem necrosis, and necrotic spots on the fruit resulting in significant yield losses in affected areas.

RT-PCR assays showed that ToTV was only present on tomato crops and *Datura stramonium* (jimson weed) samples that were collected from the Limpopo province (situated far north of South Africa). In addition, an arable weed species (*Abutilon grantii* Meeuse; appearing on SANBI's (South African National Biodiversity Institute) red list) (http://redlist.sanbi.org/species.php?species=2595-13) growing among tomato crops in the northern part of the KwaZulu Natal province tested positive for ToTV infection (TorKZN-186, KY581570). No other torradovirus species, including *Tomato marchitez virus* (ToMarV), *Tomato chocolàte virus* (ToChV), *Tomato necrotic dwarf virus* (ToNDV), and *Tomato chocolate spot virus* (ToChSV), were identified in South Africa. Additionally, all whitefly-infested symptomatic bell pepper plants that were assayed for the presence of torradovirus infections using RT-PCR were negative.

A total of 316 tomato samples, 269 bell pepper samples, and 182 weed samples belonging to six botanical families (Amaranthaceae, Asteraceae, Brassicaceae, Euphorbiaceae, Malvaceae, and Solanaceae) were tested for torradovirus infection. A high prevalence of ToTV was identified on tomato (71.6%) and D. stramonium (66.7%) in the Limpopo province of South Africa [9]. To a lesser extent, ToTV was identified on the arable weed (*Abutilon grantii*) growing among tomato crops in the northern KwaZulu Natal province (13.8%); adjacent/nearby tomato crops were not infected [9]. ToTV infection of tomatoes was restricted to the Limpopo province which also had comparably higher whitefly infestation levels. Overall, only 15.1% of tomatoes and 11% of weeds sampled from South Africa tested positive for ToTV infection [9].

Globally, ToTV has been identified in Spain, Hungary, Poland, Canary Islands, France, Panama, Italy, Australia, Colombia, and Morocco [1,10–18]. During a national survey in 2015, ToTV was identified for the first time in South Africa [9,19]. In this follow-up study, we report the first full genome sequence and phylogenetic analysis of a South African isolate of ToTV from tomato.

2. Materials and Methods

2.1. RNA Extraction

Total RNA was extracted using a Quick-RNA™ MiniPrep kit (Zymo Research, Waltham, MA, USA) according to the manufacturers' guidelines. Six hundred microliters of lysis buffer was added to 20 mg of frozen leaf tissue in a sterile 1.5 mL microcentrifuge tube containing five to six tanzanite beads. Samples were subsequently macerated using a bead beater homogenizer. Prior to the final elution step, 30 µL of nuclease-free water was added to the column and allowed to incubate for 2 min at room temperature. The quality and quantity of each RNA extract was measured using a Nanodrop 2000 spectrophotometer (Thermo Fisher Scientific Inc., Waltham, MA, USA). Samples were stored at −80 °C pending further analysis.

2.2. Reverse Transcription–Polymerase Chain Reaction (RT-PCR)

A two-step RT-PCR approach was used as the first measure to detect torradovirus infections due to the lack of commercially available antibodies. Template RNA (4 µL) was incubated at 65 °C for 5 min and subsequently kept on ice. A master mix component containing 2 µL of a gene-specific primer, 4 µL of reaction buffer, 1 µL of reverse transcriptase enzyme, 1 µL of ribolock RNase inhibitor, 2 µL of dNTPs, and 6 µL of nuclease-free water was added to make a final volume of 20 µL for each reaction. cDNA was synthesized using a RevertAid Premium Reverse Transcriptase kit (Thermo Fisher Scientific Inc., Waltham, MA, USA) according to the manufacturers' guidelines. Conditions for RT were 42 °C for 1 h and 70 °C for 10 min.

PCR was carried out in 20 µL reaction volumes using a KAPA2G Fast HotStart ReadyMix PCR kit (KAPA Biosystems, Wilmington, NC, USA). Each PCR reaction contained 10 µL of KAPA Ready Mix, 2 µL of each primer (10 ng/µL), 30 ng of template DNA, and nuclease-free water. A set of degenerate torradovirus primers Torrado-2F (corresponding to nt 2589–2608 of ToTV (PRI-ToTV0301; DQ388880); nt 2528–2547 of ToMarV (PRI-TMarV0601; EF681765); nt 2561–2580 of ToChSV (GQ305132) and nt 2568–2587 of ToChV (ToChV-G01; FJ460290)) and Torrado-2R (corresponding to nt 3084–3103 of ToTV (PRI-ToTV0301; DQ388880); nt 3023–3042 of ToMarV (PRI-TMarV0601; EF681765); nt 3056–3075 of ToChSV (GQ305132) and nt 3063–3082 of ToChV (ToChV-G01; FJ460290)) that target a 515 bp region (overlapping the Vp35 and Vp26) located on the RNA-2 strand was used to detect the presence of torradoviruses [17]. Conditions for PCR were 95 °C for 2 min; 35 cycles of 95 °C for 30 s, 51 °C for 25 s and 72 °C for 20 s followed by a final elongation at 72 °C for 10 min. PCR products were resolved on a 1.5% agarose gel stained with SYBR Safe DNA gel stain (Invitrogen, Carlsbad, CA, USA).

2.3. Cloning and Sequencing

PCR-positive amplicons were excised and purified using a Zymoclean Gel DNA Recovery Kit (Zymo Research, Irvine, CA, USA). A TA cloning kit (Invitrogen, Carlsbad, CA, USA) was used to ligate the target sequences from purified gel extracts onto a PCR 2.1 cloning vector following the manufacturers' guidelines. Chemically competent *Escherichia coli* cells were transformed by heat shock (42 °C for 30 s) prior to a 30 min incubation on ice. Successful transformants were selected using blue/white colony screening and cultured overnight at 37 °C in Luria–Bertani (LB) broth containing 50 µg/mL kanamycin. Plasmid extractions were carried out using a Zyppy Plasmid Miniprep Kit (Zymo Research, Irvine, CA, USA). The insert was confirmed using *Eco*R1 endonuclease activity (Thermo Fisher Scientific, Waltham, MA, USA). Reactions were incubated at 37 °C for 15 min and terminated at 85 °C for 5 min. Bi-directional sequencing of positive transformants were carried out at Inqaba Biotec (Pretoria, South Africa) using a 3500xL Genetic Analyzer (Applied Biosystems, Foster City, CA, USA). The Blast tool in MEGA X software [20] was used to validate the identity of each clone against sequences available on the NCBI GenBank database.

2.4. Genome Analysis

Total RNA was extracted (>50 ng/uL) from a ToTV-positive tomato (*Solanum lycopersicum* Mill.) plant collected from the Limpopo province (DMS: 23°38′22.6104″ S 30°4′40.2996″ E) and analyzed using high-throughput sequencing (HTS). HTS data were generated using an Illumina HiSeq 2500 Ultra-High-Throughput Sequencing System (Illumina Inc., Santiago, CA, USA) at the Agricultural Research Council Biotechnology Platform (ARC-BTP (Pretoria, South Africa)), and the raw data were deposited into GenBank SRA: SUB8159355. Read lengths less than 25 nucleotides were trimmed and pair-end sequence libraries were generated. The raw data were trimmed using Trimmomatics version 0.36 whereby the low-quality sequence regions and Illumina universal adapter sequences were trimmed and removed. CLC genomics workbench 9.5.3 (https://www.qiagenbioinformatics.com/) was used to remove host data by matching sequence reads to a reference tomato genome (Heinz, Accession no. NC015449) prior to de novo assembly. The remaining contigs were then identified using Blast version 2.6.0. against the NCBI nucleotide database. Contig sequences with viral identities were extracted for functional annotation with Blast2GO using the default parameters. Protein translation and ORFs were identified using ORF Finder (https://www.ncbi.nlm.nih.gov/orffinder/). A comparison of nucleotide and amino acid similarities was established using SIAS tools (http://imed.med.ucm.es/Tools/sias.html), and phylogenetic analysis of the RNA-1 and RNA-2 genome of ToTV was inferred from trees generated using a best-fit model in MEGA 6 software. The sequence of each RNA segment generated from the HTS analysis was reconstructed and verified with primers designed using SnapGene v5.1.4.1 (Tables S1 and S2). The genome of ToTV infecting tomato crops in South Africa was constructed from overlapping RT-PCR clones.

3. Results

3.1. Survey Analysis

The emergence of torradovirus-like symptoms in the Limpopo province of South Africa reduced the yield and quality of tomatoes. In severely affected crops, fruit set was suppressed (Figure 1A–C). ToTV symptoms on tomato fruit (Figure 1C) were often mistaken for *Tomato spotted wilt virus* (TSWV) by local tomato growers and government extension workers. Nearby weeds infested with whiteflies exhibited symptoms such as stunting and leaf deformation. (Figure 1D).

Figure 1. Tomato crops and nearby weed species exhibiting virus-like symptoms. (**A**): Necrotic spots beginning at the base of young leaves. (**B**): Vertical stem necrosis. (**C**): Necrotic spots and fruit deformation. (**D**): Chlorosis, necrotic spots, stunted growth, and linear chlorotic spots along the veins of *Datura stramonium*. (**E**): Stunting, chlorosis, and leaf deformation symptoms on the arable weed *Abutilon grantii* growing among tomato field crops in the northern KwaZulu Natal province.

Many of the tomato crops exhibiting torradovirus-like symptoms were concentrated in the northern parts of South Africa (particularly in the Limpopo province). Based on phenotypic analysis, both *Trialeurodes* sp. and *Bemisia* sp. were present on symptomatic field and greenhouse cultivated tomato crops. Symptom severity was heightened in areas located in the northern parts of South Africa where whitefly populations were significantly higher.

3.2. Virus Detection

Each 515 bp PCR positive product was validated by cloning and sequencing. A consensus sequence was derived from the isolates of ToTV infecting tomato crops and *D. stramonium* in the Limpopo province of South Africa based on their nucleotide sequence similarity. Blast analysis showed that the isolate of ToTV from South Africa (Lim-186, KP890356) matched 99% to the Polish isolate Wal'03 (EU563947) [19,21]. The isolate of ToTV identified on *Abutilon grantii* (family Malvaceae) growing among tomatoes in the northern KwaZulu Natal province (TorKZN-186, KY581570) was not identified on tomato crops and matched 92.8% to the isolate T795 (KX132809) from Italy [9].

3.3. Sequence Analysis

The full genome of ToTV was elucidated using high-throughput sequencing. The dataset contained 23,624,259 raw paired-end reads. A total of 127,654 contigs were generated from the de novo assembly. Only 2101 contigs did not align with the host genome and consisted of viral, bacterial, fungal, and traces of plant sequences. Of the 2101 contigs, 68 contigs were similar to known viral sequences. The longest contig obtained for the RNA-1 and RNA-2 genome of ToTV was comprised of 7420 and 5381 nucleotides (nt), respectively. Sequences generated from each RT-PCR clone (amplified with the primers listed in Tables S1 and S2) that were obtained from two individual plant extracts matched to designated regions on the RNA-1 and RNA-2 segment of ToT-186, respectively, confirming the absence of quasispecies.

3.4. RNA-1

RNA-1 (ToT-186; Accession no. MH587229) was 7420 nt in length excluding the poly (A) tail and comprised of a 109 nt 5′ untranslated region, a single open reading frame (ORF-1), and an 834 nt 3′ non-coding region (Figure 2). ORF-1 (nucleotides 110–6586) encodes a predicted 241 kDa polyprotein (6477 nt; 2158 amino acids (aa)) and contains an initiation codon (AUG), and stop codon (UGA) at positions 110–112 nt and 6584–6586 nt, respectively. According to Verbeek et al. [1], there are conserved regions in the polyprotein with motifs typically associated with a protease cofactor (PRO-co), helicase (HEL), protease (PRO) and an RNA dependent RNA polymerase (RdRP) (Figure 2). Their functional domains are PRO-co (aa 106–338), HEL (aa 337–534), 3C-like PRO (aa 1000–1100), RdRP (aa 1303–1554). The position of each domain was determined by high nucleotide (>98%) and amino acid (>99%) sequence similarities with Polish (KJ940975) and Spanish (DQ388879) isolates Table 1. Consistently higher nucleotide and amino acid sequence similarity patterns were observed among members of each torradovirus species. Interestingly, the nucleotide sequence analysis of the 3′ non-translated region showed a high level of inter and intra-specific variation among isolates. Blast analysis of the RNA-1 sequence of ToTV from South Africa (ToT-186) matched 99% to the Spanish isolate (DQ388879) and Polish isolate Kra (KJ940975).

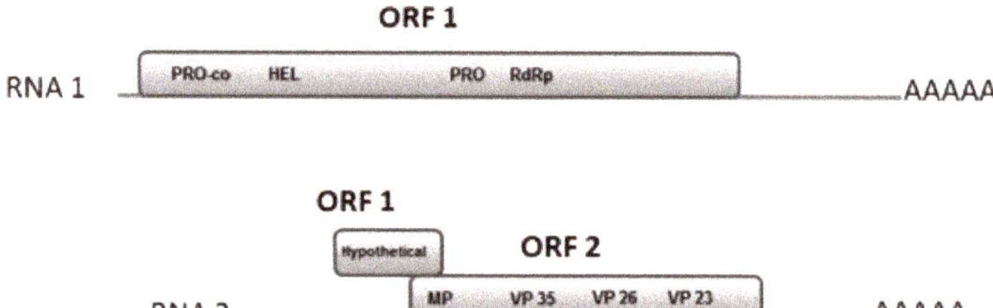

Figure 2. Bipartite genome organization of torradoviruses. RNA-1 encodes a large polyprotein (ORF-1) that contains a protease cofactor (PRO-co), helicase (HEL), 3C-like protease (PRO), and an RNA-dependent RNA polymerase (RdRp) which are involved in proteolytic cleavage and replication. RNA-2 encodes a hypothetical protein that partially overlaps ORF 2. Motifs associated with a movement protein (MP) and three coat proteins are present in ORF-2.

Table 1. A comparison of the nucleotide and amino acid similarities of coding and non-coding regions on the RNA-1 fragment of Tomato torrado virus (ToTV) from South Africa (ToT-186) with all other fully sequenced tomato-infecting torradovirus species identified throughout the world.

| Torradovirus Isolate (RNA 1) | 5'UTR | Polyprotein | ORF 1 Polyprotein | | | | | | | | | 3'UTR |
| | | | Protease co. | | Helicase | | Protease | | RdRP | | |
	nt (%)	nt (%)	aa (%)	nt (%)	aa (%)	nt (%)	aa (%)	nt (%)	aa (%)	nt (%)	aa (%)	nt (%)
ToT-186 (MH587229) South Africa	100	100	100	100	100	100	100	100	100	100	100	100
ToTV T795 (KX132808) Italy	99.06	99.05	99.72	99.71	100	98.97	100	99.33	100	98.94	100	97.82
ToTV Ros (KM091449) Poland	98.13	99.07	99.30	99.42	99.56	99.31	100	100	100	99.07	100	97.36
ToTV Wal03 (EU563948) Poland	100	98.96	99.53	99.71	100	98.97	100	99.66	100	98.94	99.60	99.50
ToTV Kra (KJ940975) Poland	100	99.10	99.49	99.57	100	98.97	100	99	100	99.07	100	99.67
ToTV (DQ388879) Spain	98.13	98.98	99.58	99.71	100	98.63	100	99.66	100	99.80	100	98.76
ToChV-G01 (FJ560489) Guatemala	72.89	38.20	30.56	N/A	N/A	24.82	14.43	N/A	N/A	72.02	87.95	61.19
ToChSV (GQ305131) Guatemala	71.02	38.01	31.86	N/A	N/A	30.30	21.21	30.34	14.92	24.20	13.09	62.05
ToMarV PRI (EF681764) Mexico	63.55	26.34	13.38	N/A	N/A	79.59	98.96	N/A	N/A	26.23	13.25	N/A
ToMarV pJL89 (KT756876) Mexico	66.35	26.53	13.42	N/A	N/A	79.59	98.96	N/A	N/A	25.97	13.25	61.24
ToMarV M (KT756874) Mexico	66.35	26.56	13.42	N/A	N/A	79.59	98.96	N/A	N/A	26.10	13.25	61.40
ToNDV R (KC999058) USA	66.35	48.56	49.83	N/A	N/A	78.91	97.93	N/A	N/A	N/A	N/A	62.24

nt—nucleotide; aa—amino acid; all values are expressed as a percentage (%); N/A—data not available. ToTV—*Tomato torrado virus*; ToChV—*Tomato chocolàte virus*; ToChSV—*Tomato chocolate spot virus*; ToMarV—*Tomato marchitez virus*; ToNDV—*Tomato necrotic dwarf virus*. Accession numbers for each isolate are listed in parenthesis.

3.5. RNA-2

RNA-2 (ToT-186; Accession no. MH587230) spanned a total of 5381 nucleotides (nt) excluding the polyadenylated tail and comprised of two ORFs, a 172 nt 5' leader sequence and a large non-coding region (1092 nt) at the 3' end (Figure 2). ORF-1 (nucleotides 173–736) encodes a 20 kDa protein (564 nt; 187 aa) with no known function or homology to other proteins in the database. ORF-2 (nucleotides 693–4289) partially overlaps ORF-1 and encodes a large 133.5 kDa polyprotein (3597 nt; 1198 aa) which includes three virion capsid subunits [21]. The coat proteins Vp35 (amino acids 483–728), Vp26 (amino acids 733–969), Vp23 (amino acids 982–1198) have a molecular weight of approximately 35 kDa, 26 kDa, and 23 kDa, respectively. A movement protein consensus sequence (LxxPxL) identified near the N-terminal region of ORF-2 indicates the likelihood of a putative movement protein [22]. Budziszewska et al. [21] established that the putative 3A movement protein (MP) encoded by ORF-2 near the N-terminal is common to both ToTV and ToMarV. Alignment of the MP indicates that it is encoded in a similar position as all tomato-infecting torradovirus species. Nucleotide and amino acid similarities of the 3A protein (Table 2) show that all ToTV species share a very high (>98% nt) and (100% aa) similarity with each other but vary by approximately 70% (nt) and 80% (aa) when compared with other torradovirus species. Interestingly, all other torradovirus species, including ToMarV, ToChV, ToChSV, and ToNDV, share equally high similarities with each other. High levels of variability are seen in the 5′ leader sequence among different torradovirus species. ToMarV and ToNDV had the highest level of nucleotide variability in the 5′ UTR (35–37% nt) when compared to ToTV isolates. All torradoviruses characteristically have a short 5′ leader sequence and an unusually long 3′ non-coding region (NCR). The 3′ NCR of ToTV extends approximately 1098 nt and ToChV more than 1400 nt. Additionally, analysis of the 3′ NCR showed that ToChV shared the lowest nucleotide similarity (<32%) with ToTV isolates. Although the percentage of nucleotide and amino acid similarities in the coat protein (CP) are conserved among torradovirus species, the VP35 showed higher levels of variability between ToTV isolates and other torradovirus species when compared to the VP26 and VP23. Blast analysis of the RNA-2 segment of ToTV from South Africa (ToT-186) matched 99% to ToTV isolates T795 (KX132809) and Ros (KM114266) from Italy and Poland, respectively.

Table 2. A comparison of the nucleotide and amino acid similarities of coding and non-coding regions on the RNA-2 fragment of ToTV from South Africa (ToT-186) with all other fully sequenced tomato-infecting torradovirus species identified throughout the world.

Torradovirus Isolate (RNA-2)				ORF-2								
	5′UTR	ORF 1		MP		VP 35		VP 26		VP 23		3′UTR
	nt (%)	nt (%)	aa (%)	nt (%)	aa (%)	nt (%)	aa (%)	nt (%)	aa (%)	nt (%)	aa (%)	nt (%)
ToT-186 (MH587230) South Africa	100	100	100	100	100	100	100	100	100	100	100	100
ToTV T795 (KX132809) Italy	97.34	99.47	98.94	99.08	100	98.78	100	99.43	100	99.38	99.53	99.53
ToTV Ros (KM114266) Poland	97.87	98.95	98.94	98.93	100	98.64	99.18	99.01	99.57	99.53	99.53	99.53
ToTV Wal03 (EU563947) Poland	97.87	99.3	99.47	98.48	100	99.05	100	99.29	100	99.07	100	99.59
ToTV Kra (KJ940974) Poland	97.34	99.47	99.47	98.63	99.08	98.5	99.59	99.15	100	99.53	99.53	99.53
ToTV (DQ388880) Spain	97.34	98.6	98.94	98.93	100	98.91	100	99.01	100	98.92	99.53	99.46
ToChV-G01 (FJ560490) Guatemala	46.27	57.41	70	70.06	80.36	63.27	73.07	73.55	07.20	65.74	81.91	31.13
ToChV-G02 (GU071087) Guatemala	46.27	60.73	71.05	71.27	80.36	63.55	73.87	73.69	86.86	65.59	80.55	31.43
ToChSV (GQ305132) Guatemala	40.42	63.52	73.68	70.36	80.82	64.9	74.69	74.68	91.52	64.36	81.94	49.23
PRI-TMarV0601 (EF681765) Mexico	36.7	60.73	72.1	70.36	80.82	64.22	76.32	73.41	89.83	65.74	79.62	45.88
ToMarV pJL89 (KT756877) Mexico	36.17	60.38	73.68	70.06	81.27	66.66	75.51	73.41	89.4	66.05	80.55	46.62
ToMarV M (KT756875) Mexico	36.17	60.55	73.68	69.9	81.27	66.53	75.51	73.55	89.4	65.89	80.55	46.62
ToNDV R (KC999059) USA	35.63	61.6	70.52	70.06	81.27	64.9	74.28	72.15	91.1	66.35	80.55	46.48

nt—nucleotide; aa—amino acid; all values are expressed as a percentage (%). ToTV—*Tomato torrado virus*; ToChV—*Tomato chocolàte virus*; ToChSV—*Tomato chocolate spot virus*; ToMarV—*Tomato marchitez virus*; ToNRV—*Tomato necrotic dwarf virus*. Accession numbers for each isolate are listed in parenthesis.

3.6. Phylogeny

Phylogenetic analysis of the full-length RNA-1 and RNA-2 nucleotide sequences showed five distinct clades representing the five known members of the torradovirus genus that are capable of infecting tomatoes (Figures 3 and 4); the taxonomy of ToChV, ToChSV, and ToNDV are presently incomplete so these viruses are not approved as distinct species (10th Report of the Internation Committee on Taxonomy of Viruses). The full-length RNA-1 genome of ToTV (ToT-186) from South Africa grouped with ToTV isolates, but it was the most evolutionary diverse. This relationship is supported by strong bootstrap values that are consistent among isolates within the ToTV clade (Figure 3). The tree topology indicates that ToTV may have originated and spread from Poland and Spain to other parts of the world including South Africa.

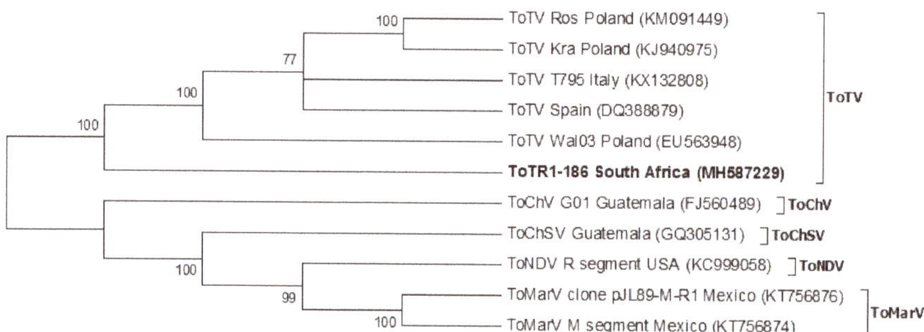

Figure 3. Phylogenetic relationship of the full-length ToTV RNA-1 genome from South Africa (ToTR1-186) with all other fully sequenced tomato-infecting torradovirus isolates to date. Evolutionary analysis was inferred using the maximum likelihood method and 1000 bootstrap replicates based on the general time reversible model [23]. A gamma distribution rate (*g* = 5) was used to model evolutionary differences among sites. A rate variation model allowed some sites to be evolutionary invariable (*I*). Accession numbers are in parenthesis.

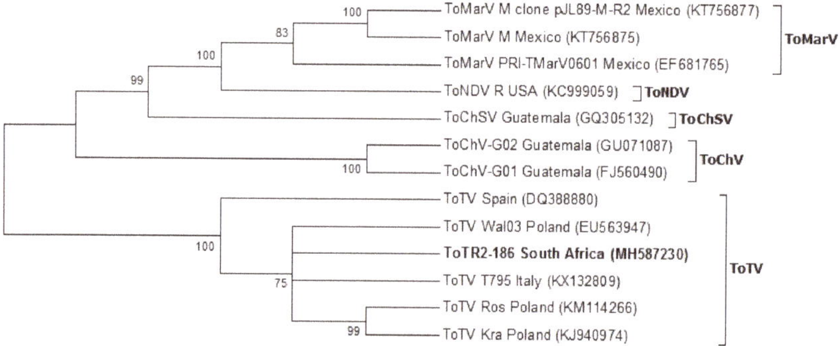

Figure 4. Phylogenetic relationship of the full-length RNA-2 genome from South Africa with all other fully sequenced tomato-infecting torradovirus isolates to date. Evolutionary analysis was carried out in MEGA 6 using the maximum likelihood method and 1000 bootstrap replicates based on the general time reversible model [23]. A gamma distribution rate (*g* = 5) was used to model the evolutionary differences among sites. Accession numbers are in parenthesis.

The full-length RNA-2 genome of ToTV from South Africa (ToT-186) did not cluster with other ToTV isolates. The tree topology in Figure 4 indicates that ToT-186 shares a closer relation to the isolates T795 from Italy and Wal'03 from Poland. The phylogram further outlines the divergence and likely spread of ToTV and other tomato-infecting torradoviruses throughout the world.

Similar tree topologies of the full-length RNA-1 and RNA-2 genomes of tomato-infecting torradoviruses are illustrated in Figures 3 and 4. ToNDV (USA) diverges from ToMarV (Mexico) followed by ToChSV (Guatemala) and finally ToChV (Guatemala). The distance of the branches indicates that ToChV is the most evolutionary diverse species when compared to those mentioned previously. Their distribution is limited to the south-western parts of the United States, Mexico, and Central America. ToTV isolates, on the other hand, form a separate group from all other torradovirus species and have been identified in parts of Europe, Australia, South America, and Africa.

Phylogenetic analysis of the torradovirus RNA-1 polyprotein showed that the isolate of ToTV (ToT-186) from South Africa clustered with ToTV Wal03 from Poland; this relationship is supported by a weak bootstrap value (Figure 5). Interestingly, ToNDV groups with ToTV isolates and clusters with the Spanish ToTV isolate (DQ388879). The Italian ToTV isolate T795 diverges from but does not group within the clade of ToTV isolates analyzed in this study. T795 also expressed the highest level of diversity among ToTV isolates based on the tree topology (Figure 5). ToChV and ToChSV clustered and formed their own clade from which a group of ToMarV isolates diverged.

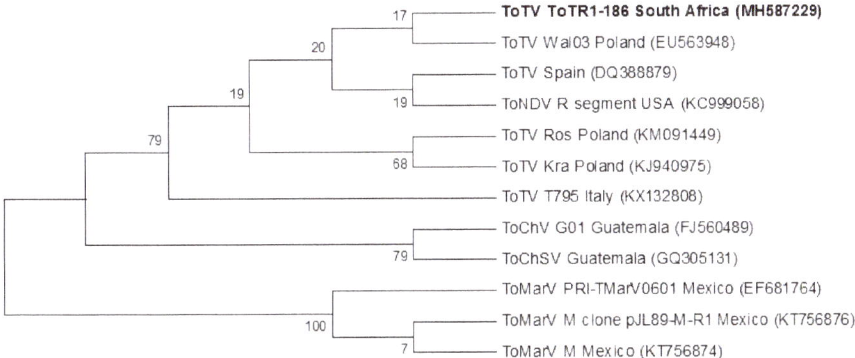

Figure 5. Phylogenetic relationship of the ToTV RNA-1 polyprotein from South Africa (ToTR1-186) in conjunction with all other tomato-infecting torradovirus isolates to date. Evolutionary analysis was inferred using the Maximum likelihood method and 1000 bootstrap replicates based on the General Time Reversible model [23]. Accession numbers are in parenthesis.

The tree topology from the phylogenetic analysis of the torradovirus RNA-2 polyprotein shows that ToTV isolates form a distinct clade from other tomato-infecting torradoviruses (Figure 6). The isolate of ToTV (ToT-186) from South Africa clustered with isolate T795 from Italy. The Spanish ToTV isolate (DQ388879) expressed the highest level of diversity and did not group with other ToTV isolates within this clade. ToNDV on the other hand, groups with ToMarV isolates and clusters with the Mexican isolate (PRI-TMarV0601). ToChSV and ToChV isolates diverge from the group of ToMarV isolates, respectively forming separate groups. These relationships are supported by strong bootstrap values (Figure 6).

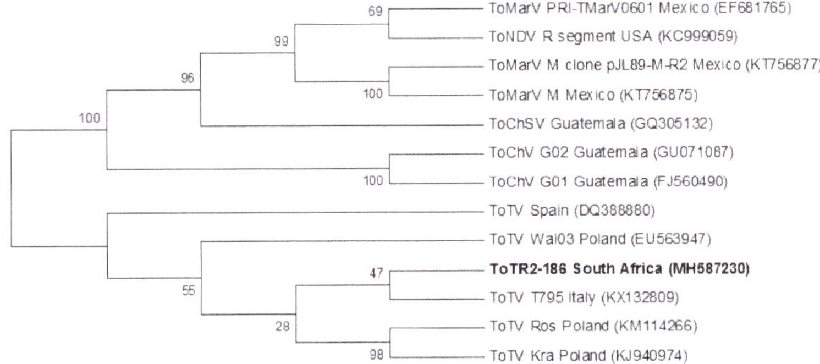

Figure 6. Phylogenetic relationship of the ToTV RNA-2 polyprotein from South Africa in conjunction with all other fully sequenced tomato-infecting torradovirus isolates to date. Evolutionary analysis was carried out in MEGA 6 using the Maximum likelihood method and 1000 bootstrap replicates based on the General Time Reversible model [23]. A gamma distribution rate ($g = 5$) was used to model the evolutionary differences among sites. Accession numbers are in parenthesis.

4. Discussion

The recent ToTV outbreak in the northern Limpopo province of South Africa caused substantial damage to fields of commercially grown tomatoes [19]. In the Limpopo province, tomatoes are cultivated on approximately 3600 hectares of farmland which accounts for half of South Africa's tomato production. Losses to the global tomato industry as a consequence of ToTV infections remain inconclusive. The presence of coinfecting viruses such as Pepino mosaic virus (PepMV) and Tomato chlorosis virus (ToCV) has hampered the efforts of researchers to acquire concise data. The findings of Gomez et al. [24] from their study on ToTV infection of tomato crops in Spain (2005–2008) found that even though most of the crops were singly infected with ToTV, symptom severity was not indifferent to mixed infections with PepMV and other viruses. In addition, they concluded that mixed infections with ToTV and PepMV modulate viral fitness and epidemiology.

In South Africa, tomato crops and weeds that tested positive for ToTV infections were often coinfected with PepMV and ToCV. These co-infecting viruses, i.e., *Potato virus Y* (PVY), PepMV and ToCV were initially identified on samples that were analyzed using high-throughput sequencing. Some tomato crops exhibited burnt-like/necrotic spot symptoms typically associated with torradovirus disease, whilst others displayed interveinal leaf chlorosis and chlorotic flecking symptoms typically associated with crinivirus infections. Symptom expression may be linked to the primary infecting virus and factors that influence viral fitness. Although evidence suggests that there are no significant associations among these viruses (ToTV + PepMV and ToTV + ToCV) [24], the symptomatology of diseased tomato crops masks the presence of coinfecting viruses and facilitates primary spread. Consequently, the dynamics of virus epidemiology are affected, and this may have negative impacts on alternative crop hosts and less tolerant varieties.

Prior to this study, whitefly-transmitted torradovirus disease was unfamiliar to South African farmers and agricultural extension workers. These symptoms were often mistaken for physiological disorders that generally result from the prolonged exposure of crops to higher temperatures or excessive pesticide applications. For many South African farmers, torradovirus infections appeared as an unrelated physiological condition and therefore requires the use of molecular assays to validate symptomatology in the field.

The symptoms associated with ToTV infections are almost indistinguishable from those of other tomato-infecting torradovirus species. Therefore, a generic set of primers [17] was used to screen for the presence of other torradoviruses. Sequence analysis of the 515 bp amplicons indicated that ToTV was the only torradovirus species infecting tomatoes and some weed species in South Africa. High-throughput sequencing of a pooled sample of all ToTV positive nucleic acid extractions confirmed the absence of other torradovirus species in South Africa. Similarly, the absence of whitefly-transmitted viruses was confirmed in symptomatic bell pepper crops infested with whiteflies. These bell peppers did not exhibit typical torradovirus symptoms. Interestingly, the emergence and distribution of ToTV on tomato crops in South Africa was restricted to the Limpopo province.

Sequence analyses of ToTV isolates from the Limpopo region infecting *D. stramonium* and tomato crops (Figure 1A–D) were similar, therefore, a sequence was selected and deposited into the NCBI nucleotide database (Lim-186, KP890356) [19]. These results indicate that the epidemiology of ToTV is influenced by the presence of weeds as alternative hosts and a source of virus inoculum. On the other hand, an isolate of ToTV (TorKZN-186, KY581570) identified from samples of the arable weed species (*Abutilon grantii*) growing among tomato field crops in the northern KwaZulu Natal region was not detected on the nearby tomato crops (Figure 1E). This isolate of ToTV may not be a tomato-infecting strain and hence did not pose a threat to tomato production in KwaZulu Natal [9]. Biological assays and host indexing are required to validate the epidemiology of this isolate.

Studies conducted by Alfaro-Fernández et al. [25] reported that ToTV outbreaks in Spain and Poland occurred as a result of high vector populations. During the emergence of ToTV in the Limpopo province of South Africa, whitefly populations reached an unprecedented level. Efforts to control the infestations did not effectively reduce the pest pressure. In addition, South Africa experienced a period of drought with soaring temperatures exceeding 35 °C (95 °F). The warm dry conditions may have influenced the dynamics of whitefly vector populations and fueled their infestation levels. The emergence of ToTV in the warmer and dryer parts of South Africa indicates that climate is an integral part of virus–vector epidemiology. Moreover, South Africa experienced the worst drought in more than 30 years with the highest recorded temperatures in history suggesting that extreme weather events driven by *El Nino* and other climate change phenomena may have contributed toward the emergence of ToTV in the Limpopo province.

ToTV is transmitted by three whitefly species belonging to the genera *Bemisia* and *Trialeurodes*. Both genera were identified in South Africa from adult and nymph phenotypic screening. On the other hand, ToTV can be transmitted mechanically [1], but mechanical inoculation assays on solanaceous hosts including *Solanum lycopersicum, Capsicum annuum, Nicotiana benthamiana, Nicotiana tabacum, Nicotiana glutinosa,* and *Solanum melongena* with isolates of ToTV from Limpopo were unsuccessful. Our results indicate that these isolates of ToTV may not be easily mechanically transmissible—a trend generally associated with viruses that are semi-persistently transmitted. Moreover, Pospieszny et al. [26] indicate that ToTV is poorly transmitted mechanically, and this may be attributed to the low stability of ToTV virions in plant sap and low accumulation of the virus in host tissue [27]. Whiteflies collected from the leaf surface of ToTV-positive samples in the Limpopo province constituted a mix of *Bemisia* sp. and *Trialeurodes* sp., but *Bemisia* sp. was more abundant in field samples and to a lesser extent in greenhouses. ToTV-positive weed samples collected from the KwaZulu Natal region were predominantly infested with *Bemisia* sp.

The emergence of ToTV in Australia raised suspicions about the possibility of seed transmission because of the strict import regulations of plant material into the country [28]. Reports of seed transmission of a Polish ToTV isolate at a rate of 0.5% to 0.8% were obtained from seeds of mechanically inoculated pepper and tomato crops [29,30]. Other isolates may show a higher or lower affinity for seed transmissibility, but this is yet to be determined [28]. The exchange of seed and plant material throughout the world is very likely the cause of outbreaks that occur in remote locations from the epicenter of the disease.

The distribution of ToTV in South Africa was restricted to the Limpopo and northern KwaZulu Natal provinces. A significantly higher incidence of ToTV was recorded from the Limpopo province [9]. Although South Africa had a low overall prevalence of ToTV [9], it is important to raise awareness of the symptoms and subsequent economic implications associated with ToTV infections on tomatoes. The distribution of tomato-infecting torradoviruses such as ToMarV, ToChV, and ToChSV has been limited to South and Central America, whereas the occurrence of ToNDV has only been reported from the USA. ToTV is by far the most widespread tomato-infecting torradovirus that has emerged in many European countries, South America, Oceania, and Africa. Recently, another group of torradoviruses known as non-tomato-infecting torradoviruses was identified on hosts such as lettuce, carrot, cassava, and motherwort [28]. The discovery of non-tomato-infecting (NTI) torradoviruses was made possible with the advent of high-throughput sequencing (HTS) technology and raises questions about the genetic diversity, mutation rate, and fitness of torradoviruses. To this end, molecular and evolutionary studies are likely to address these concerns.

The full genome of ToTV infecting tomato crops in the Limpopo province was generated using HTS technology. RNA-1 (ToT-186) from South Africa matched 99% with Spanish (DQ388879) and Polish (Kra (KJ940975)) isolates. Phylogenetic analysis of the full-length RNA-1 genome showed that ToT-186 grouped with other ToTV isolates but was also the most evolutionary diverse (Figure 3). RNA-2 (ToT 186), on the other hand, matched 99% to Italian T795 (KX132809) and Polish Ros (KM114266) isolates. Figure 4 shows that the full-length RNA-2 genome among ToTV isolates are more evolutionary diverse and may be the consequence of coevolutionary events and mutations associated with virulence. In contrast, the phylogenetic relationships among other torradoviruses were more consistent (Figures 3 and 4). The evolutionary distance of each clade indicates that ToMarV is the most distant relative of ToTV. Although only one isolate of ToNDV and ToChSV has been fully sequenced to date, they are likely to be more widespread. The genome of ToNDV was only recently characterized by Wintermantel et al. [6], who claim that it significantly damaged tomato crops in southern California as far back as 1980.

Comparisons of the torradovirus RNA-1 (Figure 5) and RNA-2 (Figure 6) polyproteins were also carried out to account for the disparity associated with varying lengths of different torradovirus isolates and the organizational variation of full-length RNA sequences. Although Figures 5 and 6 display a similar tree topology to that of Figures 3 and 4, respectively, notable variations were identified. ToNDV in particular initially diverged from a group of ToMarV isolates (Figures 3 and 4). Both of these relationships were supported by strong bootstrap values (≥99%); a comparison of the torradovirus RNA-1 polyprotein (Figure 5) showed that ToNDV not only grouped with ToTV isolates but clustered with the Spanish isolate (DQ388879). Interestingly, the RNA-2 polyprotein of ToNDV clustered with the Mexican ToMarV isolate PRI-TMarV0601 within a clade of ToMarV isolates (Figure 6), suggesting that ToNDV may be the result of genetic exchange (recombination/reassortment) between ToTV and ToMarV; no recombination events were detected using recombinant detection program 4 (RDP4) [31].

The representation of the phylogenetic trees in Figure 3 (full-length RNA-1), Figure 4 (full-length RNA-2), Figure 5 (RNA-1 polyprotein), and Figure 6 (RNA-2 polyprotein) indicates that ToTV isolates may have diverged from the other tomato-infecting torradoviruses or vice versa. The branch lengths indicate a significant amount of genetic variation between ToTV isolates and other torradovirus relatives. The results from Tables 1 and 2 in conjunction with phylogenetic analysis (Figures 3–6) demonstrate the evolutionary traits between tomato-infecting ToTV isolates and other torradoviruses analyzed in this study. The movement of ToTV in European and African countries can be inferred from the phylogenetic trees, suggesting that the isolate of ToTV infecting tomato crops in the Limpopo province of South Africa may have originated in Poland and spread from the Mediterranean regions.

Efficient control methods can be designed to contain the spread of ToTV into other provinces of South Africa. Vector control is a key management strategy for emerging viruses such as ToTV which are naturally transmitted by *Bemisia tabaci* and two *Trialeurodes* species. Their coexistence in some locations may be an adaptive or acquired trait or possibly a type of species coevolution that is not entirely suppressing or limiting to the other. It also indicates a level of heightened fitness or the ability to adapt and thrive in the face of competition, extreme weather events, and extensive pesticide use. Weeds need to be managed more effectively, considering their role as alternative hosts. Resistant tomato varieties provide effective control against ToTV infections. These genes can also be introgressed into commercially desirable germplasm via breeding programs. In addition, good crop husbandry, intensive scouting, trap crops, fine mesh nets, mineral oils, pheromones, sticky traps, and pesticides can be used in varying combinations to manage the spread of ToTV. The use of pest phenology models may serve as an early warning system for farmers, especially those who cultivate field-grown vegetables. Importantly, there needs to be a consensus among researchers, government extension workers, farmers, and policymakers so that effective strategies can be implemented based on detailed research outputs and stringent policies that regulate the movement of seed and plant material into South Africa.

5. Conclusions

In this study, the prevalence and distribution of ToTV in South Africa are reported. In addition, a comprehensive analysis of the genome of a ToTV isolate from South Africa was generated using high-throughput sequencing technology. The development and use of such technology have led to scientific breakthroughs in many areas of research and may facilitate the identification of many new and emerging diseases. Such was the case of non-tomato-infecting (NTI) torradoviruses. Identification and characterization are the first steps toward developing methods to contain and possibly mitigate diseases such as ToTV. In countries such as Poland and Hungary, ToTV has been totally eradicated.

Supplementary Materials: The following are available online at http://www.mdpi.com/1999-4915/12/10/1167/s1, Table S1: Primers used to reconstruct the RNA-1 segment of an isolate of ToTV infecting tomato crops in South Africa., Table S2: Primers used to reconstruct the RNA-2 segment of an isolate of ToTV infecting tomato crops in South Africa.

Author Contributions: Conceptualization: V.M.; methodology: V.M.; software: V.M.; validation: V.M.; formal analysis: V.M.; investigation: V.M.; resources: A.G. and P.L.M.; data curation: V.M.; writing—original draft preparation: V.M.; Writing—review and editing: V.M.; visualization: V.M.; supervision: A.G. and P.L.M.; project administration: A.G. and P.L.M.; funding acquisition: P.L.M. All authors have read and agreed to the published version of the manuscript.

Funding: National Research Foundation: 86893.

Acknowledgments: The authors would like to extend their gratitude to the National Research Foundation (NRF) of South Africa (Grant number 86893) for their financial assistance throughout this project (2014–2017). We would also like to thank Rebeng Maine for her assistance with the overlapping RT-PCR clones, and the tomato growers and agricultural extension services throughout South Africa for their hospitality and willingness to assist during our two-year survey—especially, Bertus Venter (Natuurboerdery), Johan Ras, and Lezel at ZZ2 farms Mooketsi, Limpopo Province, South Africa.

Conflicts of Interest: All authors listed in this study hereby declare that we have no conflict of interest. The sponsors had no role in the design, execution, interpretation, or writing of the study.

References

1. Verbeek, M.; Dullemans, A.M.; van den Heuvel, J.F.J.M.; Maris, P.C.; van der Vlugt, R.A.A. Identification and characterization of tomato torrado virus, a new plant picorna-like virus from tomato. *Arch. Virol.* **2007**, *152*, 881–890. [PubMed]

2. Sanfacon, H.; Iwanami, T.; Karasev, A.V.; van der Vlugt, R.; Wellink, J.; Wetzel, T.; Yoshikawa, N. "Family Secoviridae". In *Virus Taxonomy. Ninth Report of the International Committee on Taxonomy of Viruses*; King, A., Adams, M.J., Carstens, E.B., Lefkowitz, E., Eds.; Elsevier-Academic Press: Amsterdam, The Netherlands, 2011; pp. 881–899.

3. Le Gall, O.; Christian, P.; Fauquet, C.M.; King, A.M.; Knowles, N.J.; Nakashima, N.; Stanway, G.; Gorbalenya, A.E. Picornavirales, a proposed order of positive-sense single-stranded RNA viruses with a pseudo-T = 3 virion architecture. *Arch. Virol.* **2008**, *153*, 715–727.

4. Chandrasekar, V.; Johnson, J.E. The structure of tobacco ringspot virus: A link in the evolution of icosahedral capsids in the picornavirus superfamily. *Structure* **1998**, *6*, 157–171. [PubMed]

5. Verbeek, M.; van Bekkum, P.; Dullemans, A.; van der Vlugt, R.A.A. Semi-persistent and stylet-borne manner by three whitefly vectors. *Virus Res.* **2014**, 186.

6. Wintermantel, W.M.; Hladky, L.L.; Cortez, A.A. Genome sequence, host range, and whitefly transmission of the torradovirus *Tomato necrotic dwarf virus*. *Acta Hortic.* **2018**, *1207*, 295–302.

7. European and Mediterranean Plant Protection Organization (EPPO). 2009. Tomato Torrado Virus. Available online: https://www.eppo.int/QUARANTINE/Alert_List/.../Tomato_torrado_virus.docx/ (accessed on 22 January 2017).

8. Jones, R.A.C. Future scenarios for plant virus pathogens as climate change progresses. In *Advances in Virus Research*; Academic Press: Cambridge, MA, USA, 2016; Volume 95, pp. 87–147.

9. Moodley, V.; Gubba, A.; Mafongoya, P.L. A survey of whitefly-transmitted viruses on tomato crops in South Africa. *Crop. Prot.* **2019**, *123*, 21–29.

10. Alfaro-Fernández, A.; Córdoba-Sellés, C.; Cebrián, M.C.; Sánchez-Navarro, J.A.; Espino, A.; Martín, R.; Jordá, C. First report of tomato torrado virus in tomato in the Canary Islands, Spain. *Plant. Dis.* **2007**, *91*, 1060.

11. Pospieszny, H.; Borodynko, N.; Obrepalska-Steplowska, A.; Hasiow, B. The first report of tomato torrado virus in Poland. *Plant. Dis.* **2007**, *91*, 364.

12. Alfaro-Fernández, A.; Bese, G.; Córdoba-Sellés, C.; Cebrián, M.C.; Herrera-Vásquez, J.A.; Forray, A.; Jorda, C. First report of Tomato torrado virus infecting tomato in Hungary. *Plant. Dis.* **2009**, *93*, 554.

13. Herrera-Vasquez, J.A.; Alfaro-Fernández, A.; Cordoba-Selles, M.C.; Cebrian, M.C.; Font, M.I.J.C. First report of tomato torrado virus infecting tomato in single and mixed infections with cucumber mosaic virus in Panama. *Plant. Dis.* **2009**, *93*, 198.

14. Verdin, E.; Gognalons, P.; Wipf-Scheibel, C.; Bornard, I.; Ridray, G.; Schoen, L.; Lecoq, H. First report of tomato torrado virus in tomato crops in France. *Plant. Dis.* **2009**, *93*, 1352. [CrossRef]

15. Davino, S.; Bivona, L.; Iacono, G.; Davino, M. First report of Tomato torrado virus infecting tomato in Italy. *Plant. Dis.* **2010**, *94*, 1172. [CrossRef] [PubMed]

16. Gambley, C.F.; Thomas, J.E.; Persley, D.M.; Hall, B.H. First report of Tomato torrado virus on tomato from Australia. *Plant. Dis.* **2010**, *94*, 486. [PubMed]

17. Verbeek, M.; Dullemans, A. First Report of Tomato torrado virus infecting tomato in Colombia. *Plant. Dis.* **2012**, *96*, 592. [CrossRef] [PubMed]

18. European and Mediterranean Plant Protection Organization (EPPO). 2018. Available online: https://gd.eppo.int/taxon/TOTV00/distribution/MA/ (accessed on 15 April 2018).

19. Moodley, V.; Gubba, A.; Mafongoya, P.L. First Report of Tomato torrado virus on tomato (*Solanum lycopersicum*) in South Africa. *Plant. Dis.* **2016**, *100*, 231. [CrossRef]

20. Tamura, K.; Stecher, G.; Peterson, D.; Filipski, A.; Kumar, S. MEGA6: Molecular Evolutionary Genetics Analysis version 6.0. *Mol. Biol. Evol.* **2013**, *30*, 2725–2729. [CrossRef]

21. Budziszewska, M.; Obrepalska-Steplowska, A.; Wieczorek, P.; Pospieszny, H. The nucleotide sequence of a Polish isolate of Tomato torrado virus. *Virus Genes* **2008**, *37*, 400. [CrossRef]

22. Mushegian, A.R. The putative movement domain encoded by nepovirus RNA-2 is conserved in all sequenced nepoviruses. *Arch. Virol.* **1994**, *135*, 437–441.

23. Nei, M.; Kumar, S. *Molecular Evolution and Phylogenetics*; Oxford University Press: New York, NY, USA, 2000.

24. Gomez, P.; Sempere, R.N.; Amari, K.; Gomez-Aix, C.; Aranda, M.A. Epidemics of *Tomato torrado virus*, *Pepino mosaic virus* and *Tomato chlorosis virus* in tomato crops: Do mixed infections contribute to torrado disease epidemiology? *Ann. Appl. Biol.* **2010**, *156*, 401–410. [CrossRef]

25. Alfaro-Fernández, A.; Córdoba-Sellés, M.C.; Juárez, M.; Herrera-Vásquez, J.A.; Sánchez-Navarro, J.A.; Cebrián, M.C.; Font, M.I.; Jordá, C. Occurrence and geographical distribution of the 'Torrado' disease in Spain. *J. Phytopathol.* **2010**, *158*, 457–469. [CrossRef]

26. Pospieszny, H.; Budziszewska, M.; Hasiów-Jaroszewska, B.; Obrępalska-Stęplowska, A.; Borodynko, N. Biological and molecular characterization of Polish isolates of Tomato torrado virus. *J. Phytopathol.* **2010**, *158*, 56–62. [CrossRef]

27. Pospieszny, H.; Borodynko-Filas, N.; Hasiów-Jaroszewska, B.; Rymelska, N.; Elena, S.F. Transmission rate of two Polish Tomato torrado virus isolates through tomato seeds. *J. Gen. Plant. Pathol.* **2019**, *85*, 109. [CrossRef]

28. Van der Vlugt, R.A.A.; Verbeek, M.; Dullemans, A.M.; Wintermantel, W.M.; Cuellar, W.J.; Fox, A.; Thompson, J.R. Torradoviruses. *Annu. Rev. Phytopathol.* **2015**, *53*, 485–512. [CrossRef] [PubMed]

29. Pospieszny, H.; Borodynko, N.; Hasiow-Jaroszewska, B.; Rymelska, N. Seed transmission of Tomato torradovirus in tomato. In Proceedings of the 12th International Plant Virus Epidemiology Symposium, Arusha, Tanzania, 28 January–1 February 2013.

30. Pospieszny, H.; Borodynko, N. Seed transmission of *Tomato torrado virus* in pepper (*Capsicum annuum*). In Proceedings of the 16th International Congress for Molecular Plant-Microbe Interactions, Rhodes, Greece, 6–10 July 2014; p. 168.

31. Martin, D.P.; Murrell, B.; Golden, M.; Khoosal, A.; Muhire, B. RDP4: Detection and analysis of recombination patterns in virus genomes. *Virus Evol.* **2015**, *1*, vev003. [CrossRef] [PubMed]

Review

Potato Virus Y Emergence and Evolution from the Andes of South America to Become a Major Destructive Pathogen of Potato and Other Solanaceous Crops Worldwide

Lesley Torrance [1,2,*] and Michael E. Talianksy [1,3]

[1]   The James Hutton Institute, Invergowrie, Dundee DD2 5DA, UK; michael.taliansky@hutton.ac.uk
[2]   The School of Biology, University of St Andrews, North Haugh, St Andrews KY16 9ST, UK
[3]   Shemyakin-Ovchinnikov Institute of Bioorganic Chemistry of the Russian Academy of Sciences, 117997 Moscow, Russia
*   Correspondence: lesley.torrance@hutton.ac.uk

Academic Editors: Michael Goodin and Jeanmarie Verchot
Received: 15 November 2020; Accepted: 9 December 2020; Published: 12 December 2020

Abstract: The potato was introduced to Europe from the Andes of South America in the 16th century, and today it is grown worldwide; it is a nutritious staple food eaten by millions and underpins food security in many countries. Unknowingly, potato virus Y (PVY) was also introduced through trade in infected potato tubers, and it has become the most important viral pathogen of potato. Phylogenetic analysis has revealed the spread and emergence of strains of PVY, including strains causing economically important diseases in tobacco, tomato and pepper, and that the virus continues to evolve with the relatively recent emergence of new damaging recombinant strains. High-throughput, next-generation sequencing platforms provide powerful tools for detection, identification and surveillance of new PVY strains. Aphid vectors of PVY are expected to increase in incidence and abundance in a warmer climate, which will increase the risk of virus spread. Wider deployment of crop cultivars carrying virus resistance will be an important means of defence against infection. New cutting-edge biotechnological tools such as CRISPR and SIGS offer a means for rapid engineering of resistance in established cultivars. We conclude that in future, human activities and ingenuity should be brought to bear to control PVY and the emergence of new strains in key crops by increased focus on host resistance and factors driving virus evolution and spread.

Keywords: *Potyviruses*;   whole   genome   sequencing;   epidemiology;   virus   resistance; virus host interactions

---

1. Introduction

Potato virus Y (PVY) is the type species of the genus *Potyvirus*, one of the largest groups of plant viruses, containing c. 160 species [1,2]. Potyviruses are transmitted by aphids and cause economically damaging diseases in crop plants. PVY exists as several strains and has become one of the most economically important pathogens of potato and the most important virus [3], and it occurs commonly wherever potatoes are grown. The Andean region of South America is the centre of origin of potato and many wild and domesticated species grow there (Figure 1a). After the discovery and colonization of the Americas by Europeans in the 16th century, tubers of one species of domesticated potato, *Solanum tuberosum*, were taken to Europe, and from there, over time, to the rest of the world [4,5]. Potato consumption grew in popularity in the 19th century, and it became an established staple food in many countries. Potatoes are vegetatively propagated, with progeny "seed" tubers being used to establish the next crop, and, unknown at the time, potato viruses were also transported with tubers.

Potato viruses were first identified in the 1930s [6,7], and now more than 50 viruses affecting potato are known, although only a few, including PVY, cause economically important diseases [3]. Potato is the world's third most important staple food crop and an important crop supporting food security in developing countries, where potato production now exceeds that in the developed world, and viruses are major constraints on potato production systems [3,8].

Figure 1. (**a**) Potato cultivation in the high Andes of Bolivia; (**b**) symptoms of O (ordinary) strain Potato virus Y (PVY^O) on leaves of *Solanum tuberosum* Group Phureja (**c**) necrotic local lesions (indicated by white arrow heads) elicited by PVY^O in the inoculated leaves of potato cv Atlantic; (**d**) necrotic symptoms induced by N (necrotic) strain Potato virus Y (PVY^N) in tobacco leaves.

Human activity has played a major part in the spread of PVY from the South American Andes to the rest of the world, particularly through trade in plant material of unknown disease status. Since emerging from the Andean region, PVY has also become a major pathogen of tobacco and solanaceous vegetable crops [9–13]. This review highlights current knowledge of PVY population structure, epidemiology and economic impacts, mostly drawn from research on virus infections in potato. We believe that to effectively control the virus and prevent the emergence of new strains in key crops, future research should be strongly focused on host resistance and factors driving virus evolution and spread. Therefore, we also describe natural resistance mechanisms to PVY and how they can be modulated by rising temperatures under global warming. Modern biotechnology can play a role by developing genetically edited PVY-resistant crops as well as producing "vaccinated" plants by priming their antiviral defences through RNA silencing. This work is vital for the supply of nutritious food to a growing world population in disease-free and climate resilient, sustainable agricultural systems.

2. PVY Population Structure

PVY was first identified by Smith [6], and several strains infecting potato were subsequently described in the early 20th century. Five major strain groups have now been recognised [14–16]. The first strains to be recognised were the O (ordinary) (Figure 1b), N (necrotic) and C (common) strains. These strains were characterised by biological properties and symptoms in potato hosts carrying strain-specific resistance genes (hypersensitive (HR) or *N* genes) [16–18]. The O, N and C strains were distinguished using potato cultivars carrying the genes *Ny* or *Nc*, which displayed necrotic spots on leaves when inoculated with O or C, respectively (Figure 1d) [17]. The N strain induces a systemic veinal necrosis in tobacco (Figure 1c) but generally no HR and only mild or no symptoms in potato leaves, and the C strain (PVY^C) causes economically important diseases in other solanaceous crops including tobacco, tomato and pepper. In the 1980s, additional strains were reported where although mild or no symptoms were observed in leaves, some induced severe symptoms in potato tubers, such as potato tuber necrotic ringspot disease (PTNRD) [19,20]. Genome sequencing has revealed that the virus genomes of these isolates comprise sequences derived from O and N strains, these recombinant isolates included PVY^{NTN} and PVY^{N-Wi} [19,21–23]. The resistance gene *Nz* is effective against PVY^Z (shown to be the recombinant NTN). However, other recombinant strains such as PVY^{N-Wi} were not controlled by *Nz*, and they induce mild foliar symptoms with no PTNRD [18].

Phylogenetic analysis of 460 whole-genome sequences of PVY, collected from infected plants worldwide, showed that PVY originated in the Andes of South America, the centre of origin of the potato and where potatoes were first domesticated [14,15]. The date of origin (time to most recent common potyvirus ancestor) of the PVY population was estimated to be 1860 YBP. The maximum likelihood tree analysis largely supported the previous classification of strains based on host reactions, and five phylogroups were identified. Analysis of the nonrecombinant sequences produced three main lineages: the N phylogroup, which is widespread in South America and may only have spread to the rest of the world recently; the O phylogroup, isolates of which were found mostly from plants outside South America, and a branch of O, the C phylogroup, with no isolates found among South American samples. Moreover, the C isolates were often found in non-potato hosts, suggesting they may have diverged outside the Andes, possibly in Europe [15]. The analysis suggests that diversification and emergence of some of the current strains of PVY may have occurred outside South America, possibly in Europe. The first potato breeding programmes were based on a narrow genetic foundation of only a few introductions of potato, developing varieties through in-breeding and selection [4,14]. The introduction of new germplasm to combat susceptibility to late blight into potato breeding programmes in the mid-19th century probably led to PVY strain diversification, as this material also contained PVY resistance genes. Approximately half of the sequences analysed in the study by Fuentes et al. [15] comprised recombinants from N and O sequences and formed two further lineages, the R1 and R2 phylogroups. Recombinant isolates came to prominence in the 1980s and quickly spread, displacing other PVY strains in potato production systems, probably because they were uncontrolled by N genes and the mild foliar symptoms in many modern varieties enabled them to proliferate in seed crops because they escaped visual inspections [18]. Thus, since PVY was distributed from the Andes in potato tubers, the PVY population has evolved, with new strains emerging to infect non-potato hosts as well as overcome resistance of potato hosts, and human activity has been an important driver of this process [18].

3. Virus Epidemiology and Diagnostics

3.1. Natural Vectors

Inoculum level and vector abundance are the most important factors in virus spread [24], and PVY is transmitted in nature by many different species of aphids in a nonpersistent manner [9]. The virus is acquired or inoculated within seconds through aphids probing and sampling the contents of epidermal cells with their stylets. Aphids will move to other plants if they decide not to feed (transferring virus as they go) and consequently, aphids that do not colonize potato are important vectors. Winged (alate) aphids can transmit virus between crops or between plants within a crop.

Many species of aphids can experimentally transmit PVY with varying efficiencies. *Myzus persicae* and *Macrosiphum euphorbiae* are efficient vectors that colonize potato, whereas noncolonizing cereal aphids such as *Rhopalosiphum padi* and *Sitobion avenae* that migrate in large numbers during the growing season are also important vectors. PVY incidence is correlated with aphid abundance early in the growing season [24], and so aphid incidence in the environment is often monitored. In the UK, the mean temperature in winter months (January and February) is considered a reliable guide to forecast the incidence and abundance of *M. persicae* [25]. Zhou et al. [26] found that winter temperature was the dominant factor affecting the phenology of five aphid species, including *M. persicae* and *S. avenae*, and that aphid migrations are 4–19 days earlier with each 1 °C increase in average winter temperature.

Application of insecticides is generally ineffective in decreasing PVY spread as they do not work fast enough to prevent aphids probing and transmitting the virus to a healthy plant. Mineral oils have been shown to be partially effective, but they must be applied frequently to ensure new growth is protected. The oils may interfere with aphid feeding or the interaction between virus particles and aphid stylets, but they can cause phytotoxicity and leaf marking that can obscure symptoms and interfere with visual inspection for virus. However, a combination of insecticide and mineral oil sprays applied weekly over the growing season was effective in decreasing PVY spread compared with unsprayed crops, and the effect was more pronounced when input seed had low levels of PVY (<1%) [24]. In another study, application of a straw mulch also significantly decreased PVY incidence by reducing aphid landing on the crop, whereas, in comparison, spraying mineral oil gave variable results [27]. Border crops to attract migrant aphids can also be useful as any aphid attracted to feed will lose the virus charge on probing. It was shown that covering pepper plants with 2 m wide plastic rowcovers immediately after transplanting protected them from aphid-borne virus and increased marketable yield [28].

It is known that the landscape structure affects the composition of arthropod communities within it [29]. A study of landscape composition, in terms of the relative amounts of cropland and unmanaged land, was conducted to investigate the effect on the incidence of PVY in potato crops [29]. They showed that more virus was found on farms located in areas with more crop land cover, whereas those farms in complex (unmanaged or nonagricultural) landscapes had much less virus (c. 30% vs. negligible, respectively). The results indicate that isolating potato production from other crops would decrease virus incidence.

3.2. Virus Sources

Common sources of infection are virus-infected potato plants, which can be infected tubers in the seed used to establish the crop, groundkeepers in neighbouring crops growing from tubers left in the soil or neighbouring crops being grown for consumption, which tend to have higher levels of virus than seed crops. Solanaceous weeds such as nightshades can also be reservoirs of infection [30]. PVY infection usually does not kill the susceptible plant outright, but virus will infect the progeny tubers, which in turn pass infection to the plants growing from these tubers; thus, infection quickly builds up in the tuber stock over subsequent field generations. In the most advanced potato production systems, classification programmes are in place to certify virus levels in seed stocks. These stipulate maximum virus levels, with the highest-grade seed having zero or a very small percentage of virus infection [18,31]. Virus levels in these stocks are usually estimated by visual inspections during the growing season. PVY strains that induce very mild or no symptoms in foliage can therefore be missed by visual inspection, and this has allowed virus levels to increase [18].

In the last 20–30 years, the recombinant strains of PVY have displaced the classical O and N strains and become the prevalent strains in European and US potato production systems [18,32,33]. This change in PVY strain composition is thought to have occurred because some commonly grown potato cultivars are tolerant to infection or contain strain-specific N resistance genes such as Ny (which was previously effective in controlling PVY^O, the predominant strain infecting potato for many years) but are ineffective against recombinant strains. PVY^{NTN} was shown to overcome a type of host resistance that develops later in the growing season, called mature plant resistance (MPR) [34–36], infecting cv Maris Piper at the flowering stage when they were not susceptible to PVY^O [37] and indicating that PVY^{NTN} is capable of infecting plants later in the growing season. This is important in northerly countries such as Scotland that rely on plants developing MPR before aphid numbers increase.

3.3. Diagnostics

PVY can be detected by serological and RNA-based techniques in samples of potato leaves or tubers [38]. Some potato seed classification schemes employ such laboratory-based tests to support visual inspections where symptoms are not obvious. These tests reveal the presence of known viruses. More recently, next-generation sequencing (NGS) techniques have been developed that can be applied to reveal unknown viruses and multiple infections in plants (the plant virome) [39]. Such NGS techniques can be deployed to sequence large numbers of virus genomes and compile phylogenies to discover recombinant genomes and monitor mutations and the incidence of new viruses, as well as monitor spread and genome changes in response to the deployment of disease-resistant cultivars. Obtaining information on the viromes of cultivated species and wild relatives at ecosystem boundaries can be used to monitor transmission from wild to cultivated hosts and to support predictive modelling tools and to evaluate land management practices [40]. NGS techniques provide valuable tools to monitor PVY strain emergence and evolution.

4. Economic Impacts of PVY

Estimating yield losses in potato due to PVY is complicated by several factors. For example, PVY can cause more severe symptoms in mixed infections with other viruses such as PVX. In addition, the time of infection can influence severity of disease, with young plants being more susceptible and displaying more severe symptoms when infected early in the season compared with plants where MPR has become established. The largest losses are when a crop is grown from infected seed tubers, and yield losses of 30%–64% have been reported [31,41].

In experiments with three potato cultivars, Russet Norkotah, Russet Burbank and Shepody, assessing yield from crops grown from seed lots containing different levels of PVY, it was found that for every 1% of PVY in the seed, the yield was decreased by 0.18 t/ha [42] and PVY decreased marketable yield and tuber size. An economic assessment estimated losses in the state of Idaho, USA, which produces approx. 7.1 M tonnes of potato, annually valued at $1 bn, to be $34 M (direct and indirect costs of production) [43]. It was estimated that 10% PVY infection in seed could decrease returns by $90–120 per acre depending on the market sector.

PVY also causes major losses in peppers, tomato and tobacco [13]. For example, studies of the impact of PVY on field grown bell peppers showed a reduction of between 20% and 70% in the number of fruits per plant and the marketable yield depending on time of infection, with early infection most severely affecting the crop, and the marketable fruit yield was reduced up to 90% in mixed infections with CMV [44].

Potato production in low- and middle-income countries such as those in sub-Saharan Africa (SSA) is increasing, but the yields obtained are well below potential [45,46]. In SSA, production is confined to the rain-fed, cooler highland regions (>1500 m.a.s.l. (metres above sea level)) where potatoes are grown twice a year with little or no rotation; this leads to soil degradation and the build-up of soil-borne pests and diseases such as bacteria and nematodes. Farmers resort to cutting down forests to bring new land into cultivation to obtain better yields. Pest and disease surveys in the main potato-growing areas of Kenya have revealed that aphid vectors are abundant in the main growing seasons and that PVY (one of six viruses detected) is among the most predominant, being widespread in both seed and ware crops; recombinant strains PVY[NTN] and PVY[N-Wi] were also identified [47]. Although the potato yield gap is due to many factors, a major contributing factor is poor-quality seed and lack of access to seed tubers free of viruses [45,46,48]. Moreover, there was shown to be potential for a threefold increase in yield without expanding the potato production area [45]. Given the very high pest and disease pressure in the region, deployment of virus-resistant cultivars is vital to help control virus spread and improve productivity.

5. Host Resistance and Susceptibility

Breeding for virus resistance has not been a priority because seed certification systems were very effective in controlling virus. However, the mild foliar symptoms induced by some recombinant isolates have allowed these strains to avoid detection, meaning seed programmes based on visual inspection during the growing season are much less effective. Furthermore, given the problems and cost of PVY management through ineffective vector control measures, and with the incidence of vector aphids likely to increase due to increased survival over warmer winters, the deployment of resistant cultivars is considered to be the most effective and efficient means to control PVY [3].

Plant resistance to viruses is multifaceted. In the case of incompatible interactions, two main types of dominant host resistance against PVY are present in potato: strain-specific hypersensitive response (HR) or programmed cell death conferred by the *Ny* genes (as noted above) and extreme resistance (ER), conferred by the *Ry* genes. Potato plants containing various *Ry* genes are immune to PVY. ER is typically effective against a broad spectrum of PVY strains, and virus infection upon ER is limited to a few epidermal cells. Sources of *Ry* genes from several potato species are known [31,49,50]. To date, these ER genes have provided durable resistance against PVY.

The complex molecular mechanisms of HR and ER against PVY are described in detail in Baebler et al. [51]. Briefly, a first layer of innate immunity against viruses (as well as against other pathogens) is the recognition of pathogen-associated molecular patterns (PAMPs) by pattern recognition receptors (PRRs), which leads to PAMP-triggered immunity (PTI), where viral dsRNA plays the role of possible PAMPs [52]. A second layer of immune response occurs in plants carrying resistance (*R* or *N*) genes that employ effector-triggered immunity (ETI). In the case of HR, this involves the interaction between virus-derived effectors and host resistance R or N (mostly nucleotide-binding and leucine-rich repeat (NB-LRR) proteins that trigger a number of intracellular signalling events, which lead to disease resistance [51,53]. ER-mediated mechanisms are believed to be quite similar to those of HR. However, the ER conferred by the NB-LRR protein encoded by *Ry*$_{sto}$ does not depend on salicylic acid (SA) and requires EDS1 and NRG1 proteins. Moreover, in contrast to most HR-related resistance genes, *Ry*$_{sto}$ is not temperature sensitive [54].

In compatible interactions of potato with various viruses and PVY in particular, PTI-based SA-mediated signalling pathways also play important roles in determining resistance [55–58].

Another critical factor contributing to effective virus resistance is RNA interference (RNAi) or RNA silencing. RNAi is a sequence-specific mechanism degrading foreign nucleic acids and regulating endogenous gene expression. RNAi-based defence responses involve degradation of virus-derived double-stranded RNAs (dsRNAs) into small interfering RNAs (siRNAs), which in a complex with some plant proteins mediate the sequence-specific inactivation/degradation of viral RNAs [59–61]. However, during evolution, viruses have developed mechanisms to fight back against RNAi. For example, members of the genus *Potyvirus*, and PVY in particular, encode a silencing suppressor, which is the Helper Component—Proteinase HC-Pro protein [62]. Thus, the outcome of the PVY infection may be determined by a race between the activities of RNAi and silencing suppression.

Remarkably, plant–virus interactions may also be regulated via the interplay between virus accumulation and plant methylation cycles (MTC) [63,64]. The MTC is functionally related to RNAi-based mechanisms, in which siRNAs (as major components of the RNAi-mediated defence response) are stabilized by MTC-associated transmethylation [65]. Plant DNA methylation, an epigenetic mechanism triggered by the MTC, has also been suggested to play an important role in modulating host responses to viruses by modifying functions of host genes and affecting gene expression [66,67]. Another factor released as a product of the MTC is ethylene, a phytohormone that, like other plant hormones, plays essential roles in plant responses to plant viruses [68]. Moreover, the potyviral HC-Pro silencing suppressor has been shown to physically interact with some components of the MTC [69], further confirming functional links between RNA silencing and MTC activities.

Another form of host resistance is recessive resistance, for example, that conferred by the naturally existing resistant isoforms of eukaryotic translation initiation factors, eIF4E and eIFiso4E, has been shown to inhibit virus replication. Recessive resistance has been shown to be effective against PVY in vegetable crops such as tomato and pepper [70,71] and potato [72].

Finally, it is worth noting that some plant host factors may not only provide resistance against viruses but also may be hijacked by viruses for their own benefits [73]. For example, a key "signature" component of subnuclear Cajal bodies, coilin, is recruited by PVY to increase virus pathogenicity [73].

Thus, to survive in nature, PVY, like other plant viruses, has evolved virulence strategies to overcome host defences. The co-evolutionary arms race between PVY and the host has shaped current multifactorial defence and counter-defence mechanisms. Interestingly, several studies indicate the existence of inter-relationships between various antiviral defence (resistance) mechanisms. SA has been demonstrated to induce RDR1, which is a component of the antiviral RNAi pathway [74]. Moreover, some viral silencing suppressors may modulate SA-mediated resistance to several viruses, further confirming the existence of cross-talk between SA-mediated signalling and RNAi [75–77]. MTC-based defence pathways are also tightly inter-related with RNAi defence (through stabilisation of siRNAs, as noted above) and SA-mediated response (possibly through effect on the accumulation of another phytohormone, ethylene; see above). Thus, it is conceivable that different mechanisms of plant virus interactions (defence and counter-defence) form a specific integrated system that determines susceptibility/resistance to PVY in potato.

6. Effect of Environmental Stress: Temperature

Usually, potato, like other crops, is simultaneously exposed to various stresses, which modulate plant–virus interactions, and may cause further reductions in crop yield. Under climate change, temperature appears to be one of the critical environmental factors affecting plant growth and productivity. Potato is a cool-weather crop with optimal growth at temperatures ranging between 14 and 22 °C, above these temperatures, tuber yield is dramatically decreased [78]. From climate models it is expected that heat stress impacts on potato plants will become increasingly common, with potentially damaging effects on potato production over the world [79]. To cope with high temperatures, plants have evolved a variety of mitigation strategies that facilitate thermotolerance (for example, reference [80]).

It is also well established that heat stress can significantly but differentially affect plant–pathogen interactions via the modulation of host defence responses [58,81]. With regards to PVY, higher temperatures may result in positive or negative effect on the virus replication and spread. In incompatible interactions, most resistance genes, such as *Ny* from *S. sparsipilum* and *S. sucrense*, and *Ny-1* in potato cv. Rywal, confer resistance only at cooler temperatures (16–20 °C); at higher temperatures (24–28 °C) resistance does not develop, and PVY spreads systemically throughout the plant. In contrast, resistance to PVY^N expressed in *S. stoloniferum* (Ry_{sto}) and *S. chacoense* (Ry_{chc}) is effective at both low and elevated temperatures [82,83].

A differential impact of temperature on PVY accumulation has also been observed in compatible interactions: in cv. Chicago, elevated temperatures significantly increased susceptibility to PVY, whereas the effect of heat stress in cv. Gala was negligible [58]. However, mechanisms underlying thermo-sensitivity of defence responses in incompatible and compatible interactions are different. In incompatible interactions, temperature-dependent defence is seemingly due to temperature-sensitive conformational loss of function [84] in most resistance proteins, although products of Ry_{sto} and possibly Ry_{chc} resistance genes seem to be temperature-resistant and do not lose their activity at elevated temperatures [54]. In contrast, in compatible interactions, temperature-sensitive response might be controlled by the impact of heat stress on other regulatory components of the integrated defence system.

Several effects of temperature on non-*R*- or *N*-gene-based host response mechanisms have been observed. First, it has been reported that RNAi-based defence is promoted by elevated temperatures, which may concomitantly attenuate development of the virus disease [85]. In contrast, RNAi suppression activity of the PVY suppressor HC-Pro is downregulated by higher temperatures, which could decrease defence and enhance PVY infection [86]. Second, activity of another anti-PVY defence factor, MTC, is significantly perturbed by rising temperatures in potato cv Chicago [64], resulting in a burst of the PVY infection. Third, SA is involved in both antiviral defence response and the regulation of heat shock protein (HSP) production and heat stress tolerance [87]. HSPs are known to take part in virus replication [88,89]. In turn, virus infections can modulate accumulation of HSPs [90].

Altogether, these findings suggest that responses to PVY infection and heat stress in potato have some common underlying mechanisms, which can be integrated in a network. Particular components of these networks may dominate in different virus–plant/cultivar combinations, allowing the defence responses under heat stress to be fine-tuned in a cultivar-specific manner.

7. Engineering PVY Resistance in Potato

Genetic improvement for PVY resistance is essential for sustainable potato production. Conventional breeding to incorporate major resistance genes is still a useful approach to develop new cultivars, but the extreme heterozygosity and complex genetics of potato mean that even with new genetic marker technologies, it is a time-consuming and laborious process [4,91]. Another well-established approach to develop PVY resistance is based on transgenic technology. Transgenic potato plants overexpressing PVY-derived coat protein, PVY-specific dsRNA (for RNAi) or modified plant eIF4E all demonstrated a high level of resistance. However, the commercial development of transgenic potato or other vegetable crops is constrained by regulations surrounding the release of GM plants and negative public perception. Both these approaches have been extensively discussed in a number of previous reviews [92,93]. In this review, we will focus on two technologies that have emerged in the past decade, namely CRISPR/Cas (clustered regularly interspaced short palindromic repeats/CRISPR-associated genes) and spray-induced gene silencing (SIGS), which provide new methods for improvement of PVY resistance that may be less contentious.

CRISPR/Cas is a prokaryotic adaptive immune system that has been reprogrammed into a precise and powerful tool for precise gene targeting [94]. In this system, Cas9 DNA exonuclease is guided by a short RNA (sgRNA) that defines the genomic DNA target to be modified (inducing deletions, insertions or replacements). Other types of exonucleases such as Cas13 or FnCas9 can target RNA molecules. The CRISPR/Cas system has now been extensively exploited to generate plant virus resistance. This has been achieved either by direct inhibition of viral RNAs/DNAs or by introduction of mutations into host plant "susceptibility" genes [95]. Both approaches have been successfully used to derive resistance to PVY. In the first approach, Zhan et al. [96] engineered resistance by directly targeting the PVY *P3*, *CI*, *NIb* and *CP* genes in transgenic potato expressing Cas13 and gene-specific sgRNAs.

In another approach, Makhotenko et al. [97] used the CRISPR/Cas9 tool to generate PVY resistance by targeted mutagenesis of the coilin gene in potato. An important aspect of this work is that the authors developed a new technology to achieve transgene-free genome edits and avoid the use of DNA at all. For this purpose, they delivered DNA-free CRISPR/Cas9 RNP complex pre-assembled from Cas9 and sgRNA into apical meristematic tissues of potato.

Spray-induced gene silencing (SIGS) is another RNAi-based genome technology for targeting various pests and pathogens (including viruses). Application of exogenous dsRNAs by spraying plants has been successfully exploited to induce resistance to different viruses in a wide range of crops [95]. SIGS technology for disease control appears to be potentially sustainable and environmentally friendly and could be used to protect potato from PVY.

8. Conclusions and Perspectives

Human activity has been responsible for the spread of PVY, with consequent severe losses in yields of potato and other solanaceous crops. PVY phylogeny suggests that the PVY population continues to evolve, with new strains emerging that infect new, non-potato hosts as well as defeating host resistance responses. New NGS technologies provide a powerful means to support modelling to identify the emergence and spread of new strains. PVY is transmitted by many species of aphids and can be acquired and transferred from plant to plant in seconds by the aphid stylet probing of plant cells. Aphid vector control methods are only partially effective, and the widespread use of agri-chemicals for their control is environmentally undesirable. Aphid populations are predicted to increase in size and migrate earlier in the growing seasons in warming environments, increasing the risk of virus spread. Therefore, host resistance is concluded to be the most economically effective and efficient means of control in sustainable potato production systems. Some of the known PVY resistance genes can be ineffective in warmer environments; recent research has produced much knowledge of virus–host interactions in response to abiotic stresses to aid understanding of host resistance mechanisms, however, more work is needed on how virus resistance is affected by temperature. New, advanced biotechnological tools such as CRISPR and SIGS offer huge potential to introduce virus resistance into established cultivars, thus enabling rapid development and deployment of these enhanced cultivars for efficient and sustainable crop production. However, gene target identification is an important challenging step. Another challenge is to develop efficient technologies for delivery of biomolecules such as dsRNA into intact plant cells. The production of varieties using long-established transgenic technology is contentious in some countries; the new CRISPR and SIGS technologies have the advantage of enabling modifications without the introduction of additional transgene DNA so may be more publicly acceptable, however, this remains to be resolved.

Funding: This research was funded by the Scottish Government: Rural and Environment Science and Analytical Services Division (M.T. and L.T.), and BBSRC grants BB/L011840/1, BB/N023293/1, BB/P022553/1 to L.T. and the Government of Russian Federation Grant No. 14.W03.31.0003 to MT

Acknowledgments: We thank Graham Cowan for assistance in preparing the figure.

References

1. Wylie, S.J.; Adams, M.; Chalam, C.; Kreuze, J.; López-Moya, J.J.; Ohshima, K.; Praveen, S.; Rabenstein, F.; Stenger, D.; Wang, A.; et al. ICTV Virus Taxonomy Profile: Potyviridae. *J. Gen. Virol.* **2017**, *98*, 352. [CrossRef] [PubMed]
2. ICTV Report on Virus Classification and Taxon Nomenclature. Positive-Sense RNA Viruses > Potyviridae. Genus Potyvirus. Available online: https://talk.ictvonline.org/ictv-reports/ictv_online_report/positive-sense-rna-viruses/w/potyviridae/572/genus-potyvirus (accessed on 13 November 2020).
3. Kreuze, J.F.; Souza-Dias, J.A.C.; Jeevalatha, A.; Figueira, A.R.; Valkonen, J.P.T.; Jones, R.A.C. Viral Diseases in Potato. In *The Potato Crop: Its Agricultural, Nutritional and Social Contribution to Humankind*; Campos, H., Ortiz, O., Eds.; Springer: Berlin/Heidelberg, Germany, 2020; pp. 389–430.
4. Bradshaw, J.E. Potato-breeding strategy. In *Potato Biology and Biotechnology: Advances and Perspectives*; Vreugdenhil, D., Ed.; Elsevier: Amsterdam, The Netherlands, 2007; pp. 157–177.
5. Gibbs, A.J.; Hajizadeh, M.; Ohshima, K.; Jones, R.A.C. The Potyviruses: An Evolutionary Synthesis Is Emerging. *Viruses* **2020**, *12*, 132. [CrossRef]

6. Smith, K.M. On the composite nature of certain potato virus diseases of the mosaic group as revealed by the use of plant indicators and selective methods of transmission. *Proc. R. Soc. Lond.* **1931**, *109*, 251–266.

7. Salaman, R.N. Crinkle 'A' an infectious disease of the potato. *Proc. R. Soc. B* **1930**. [CrossRef]

8. Devaux, A.; Goffart, J.-P.; Petsakos, A.; Kromann, P.; Gatto, M.; Okello, J.; Suarez, V.; Hareau, G. Global Food Security, contributions form sustainable potato Agri-food systems. In *The Potato Crop: Its Agricultural, Nutritional and Social Contribution to Humankind*; Campos, H., Ortiz, O., Eds.; Springer: Berlin/Heidelberg, Germany, 2020; pp. 3–36.

9. Kerlan, C.; Moury, B. Potato virus, Y. In *Encyclopedia of Virology*, 3rd ed.; Mahy, B.M.J., van Regenmortel, M.H.V., Eds.; Elsevier: Amsterdam, The Netherlands, 2008.

10. Scholthof, K.B.; Adkins, S.; Czosnek, H.; Palukaitis, P.; Jacquot, E.; Hohn, T.; Hohn, B.; Saunders, K.; Candresse, T.; Ahlquist, P.; et al. Top 10 plant viruses in molecular plant pathology. *Mol. Plant Pathol.* **2011**, *12*, 938–954. [CrossRef] [PubMed]

11. Romero, A.; Blanco-Urgoiti, B.; Soto, M.J.; Fereres, A.; Ponz, F. Characterization of typical pepper-isolates of PVY reveals multiple pathotypes within 381 a single genetic strain. *Virus Res.* **2001**, *79*, 71–80. [CrossRef]

12. Morel, C.; Gognalons, P.; Guilbaud, L.; Caranta, C.; Gebre-Selassie, K.; Marchoux, G.; Jacquemond, M. Biological and molecular characterization of two tomato strains of potato virus Y (PVY). *Acta Physiol. Plant* **2000**, *22*, 336–343. [CrossRef]

13. Latorre, B.A.; Flores, V.; Marholz, G. Effect of potato virus Y on growth, yield, and chemical composition of flue-cured tobacco in Chile. *Plant Dis.* **1984**, *68*, 884–886. [CrossRef]

14. Gibbs, A.J.; Ohshima, K.; Yasaka, R.; Mohammadi, M.; Gibbs, M.J.; Jones, R.A.C. The phylogenetics of the global population of potato virus Y and its necrogenic recombinants. *Virus Evol.* **2017**, *3*, 1–12. [CrossRef]

15. Fuentes, S.; Jones, R.A.C.; Matsuoka, H.; Ohshima, K.; Kreuze, J.; Gibbs, A.J. Potato virus, Y.; the Andean connection. *Virus Evol.* **2019**, *5*, vez037. [CrossRef]

16. Jones, R.A.C.; Vincent, S.J. Strain-specific hypersensitive and extreme resistance phenotypes elicited by potato virus Y among 39 potato cultivars released in three world regions over a 117 Yyar period'. *Plant Dis.* **2018**, *102*, 185–196. [CrossRef] [PubMed]

17. Jones, R.A.C. Strain group specific and virus specific hypersensitive reactions to infection with potyviruses in potato cultivars'. *Ann. Appl. Biol.* **1990**, *117*, 93–105. [CrossRef]

18. Gray, S.; Power, A.G. Anthropogenic influences on emergence of vector-borne plant viruses: The persistent problem of potato virus Y. *Curr. Opin. Virol.* **2018**, *33*, 177–183. [CrossRef] [PubMed]

19. Gray, S.; De Boer, S.H.; Lorenzen, J.; Karasev, A.V.; Whitworth, J.; Nolte, P.; Singh, R.; Boucher, A.; Xu, H. Potato virus Y: An evolving concern for potato crops in the United States and Canada. *Plant Dis.* **2010**, *94*, 1384–1397. [CrossRef]

20. Barker, H.; McGeachy, K.; Toplak, N.; Gruden, K.; Žel, J.; Browning, I. Comparison of genome sequence of PVY isolates with biological properties. *Am. J. Potato Res.* **2009**, *86*, 227–238. [CrossRef]

21. Glais, L.; Tribodet, M.; Kerlan, C. Genomic variability in Potato potyvirus Y (PVY): Evidence that PVY(N)W and PVY(NTN) variants are single to multiple recombinants between PVY(O) and PVY(N) isolates. *Arch. Virol.* **2002**, *147*, 363–378. [CrossRef]

22. Green, K.J.; Brown, C.J.; Gray, S.M.; Karasev, A.V. Phylogenetic study of recombinant strains of Potato virus Y. *Virology* **2017**, *507*, 40–52. [CrossRef]

23. Karasev, A.V.; Hu, X.J.; Brown, C.J.; Kerlan, C.; Nikolaeva, O.V.; Crosslin, J.M.; Gray, S.M. Genetic diversity of the ordinary strain of *Potato virus Y* (PVY) and origin of recombinant PVY strains. *Phytopathology* **2011**, *101*, 778–785. [CrossRef]

24. MacKenzie, T.D.B.; Nie, X.; Singh, M. Crop Management Practices and Reduction of On-Farm Spread of *Potato virus Y*: A 5-Year Study in Commercial Potato Fields in New Brunswick, Canada. *Am. J. Potato Res.* **2016**, *93*, 552–563. [CrossRef]

25. AHDB Aphid Forecasts. Available online: https://ahdb.org.uk/aphid-forecasts (accessed on 13 November 2020).

26. Zhou, X.-L.; Harrington, R.; Woiwod, I.P.; Perry, J.N.; Bale, J.S.; Clark, S.J. Effects of temperature on aphid phenology. *Glob. Chang. Biol.* **1995**, *1*, 303–313. [CrossRef]

27. Kirchner, S.M.; Hiltunen, L.H.; Santala, J.; Doring, T.F.; Ketola, J.; Kankaala, A.; Virtanen, E.; Valkonen, J.P.T. Comparison of straw mulch, insecticides, mineral oil, and birch extract for control of transmission of potato virus Y in seed potato crops. *Potato Res.* **2014**, *57*, 59–75. [CrossRef]

28. Avilla, C.; Collar, J.L.; Duque, M.; Pérez, P.; Fereres, A. Impact of floating row covers on bell pepper yield and virus incidence. *HortScience* **1997**, *32*, 882–883. [CrossRef]

29. Claflin, S.B.; Jones, L.E.; Thaler, J.S.; Power, A.G. Crop-dominated landscapes have higher vector-borne plant virus prevalence. *J. Appl. Ecol.* **2017**, *54*, 1190–1198. [CrossRef]

30. Cervantes, F.A.; Alvarez, J.M. Within plant distribution of Potato Virus Y in hairy nightshade (*Solanum sarrachoides*): An inoculum source affecting PVY aphid transmission. *Virus Res.* **2011**, *159*, 194–200. [CrossRef]

31. Valkonen, J.P.T. Viruses: Economical Losses and Biotechnological Potential. In *Potato Biology and Biotechnology Advances and Perspectives*; Vreugdenhil, D., Ed.; Elsevier Ltd.: Amsterdam, The Netherlands, 2007.

32. Funke, C.N.; Nikolaeva, O.V.; Green, K.J.; Tran, L.T.; Chikh-Ali, M.; Quintero-Ferrer, A.; Cating, R.A.; Frost, K.E.; Hamm, P.B.; Olsen, N.; et al. Strain-Specific Resistance to Potato virus Y (PVY) in Potato and Its Effect on the Relative Abundance of PVY Strains in Commercial Potato Fields. *Plant Dis.* **2017**, *101*, 20–28. [CrossRef]

33. Lindner, K.; Trautwein, F.; Kellermann, A.; Bauch, G. Potato virus Y (PVY) in seed potato certification. *J. Plant Dis. Prot.* **2015**, *122*, 109–119. [CrossRef]

34. Sigvald, R. Mature-plant resistance of potato plants against potato virus Y^O (PVY^O). *Potato Res.* **1985**, *28*, 135–143. [CrossRef]

35. Beemster, A.B.R. Virus translocation and mature-plant resistance in potato plants. In *Viruses of Potatoes and Seed-Potato Production*; de Bokx, J.A., van der Want, J.P.H., Eds.; Pudoc: Wageningen, The Netherlands, 1987.

36. Gibson, R.W. The development of mature plant resistance in four potato cultivars against aphid-inoculated potato virus Y^O and Y^N in four potato cultivars. *Potato Res.* **1991**, *34*, 205–210. [CrossRef]

37. Kumar, P.; Roberts, A.G.; Torrance, L. *Recombinant Strains of Potato Virus Y Overcome Mature Plant Resistance in Potato (Solanum tuberosum)*; James Hutton Institute: Dundee, UK, 2020; in press.

38. Fox, A.; Evans, F.; Browning, I. Direct tuber testing for Potato virus Y by real-time RT-PCR and ELISA: Reliable options for post harvest testing? *Bull. OEPP/EPPO Bull.* **2005**, *35*, 93–97. [CrossRef]

39. Jones, S.; Baizan-Edge, A.; MacFarlane, S.; Torrance, L. Viral Diagnostics in Plants Using Next Generation Sequencing: Computational Analysis in Practice. *Front. Plant Sci.* **2017**, *8*, 1770. [CrossRef]

40. McCann, H. Skirmish or war: The emergence of agricultural plant pathogens. *Curr. Opin. Plant Biol.* **2020**, *56*, 147–152. [CrossRef]

41. Whitworth, J.L.; Nolte, P.; McIntosh, C.; Davidson, R. Effect of Potato virus Y on yield of three potato cultivars grown under different nitrogen levels. *Plant Dis.* **2006**, *90*, 73–76. [CrossRef]

42. Nolte, P.; Whitworth, J.L.; Thornton, M.K.; McIntosh, C.S. Effect of seedborne Potato virus Y on performance of Russet Burbank, Russet Norkotah, and Shepody potato. *Plant Dis.* **2004**, *88*, 248–252. [CrossRef]

43. McIntosh, C.S. On the Economics of PVY. 2014. Available online: https://www.uidaho.edu/-/media/UIdaho-Responsive/Files/cals/programs/potatoes/proceedings/2014/McIntosh-Potato-Conference-2014.pdf?la=en&hash=584D1CB4EB988F093D5F08E09E6A1D42DB29D834 (accessed on 13 November 2020).

44. Avilla, C.; Collar, J.L.; Duque, M.; Pérez, P.; Fereres, A. Yield of bell pepper (*Capsicum annuum*) inoculated with CMV and/or PVY at different time intervals. *J. Plant Dis. Prot.* **1997**, *104*, 1–8.

45. Harahagazwe, D.; Condori, B.; Barreda, C.; Bararyenya, A.; Byarugaba, A.A.; Kude, D.A.; Lung'aho, C.; Martinho, C.; Mbiri, D.; Nasona, B.; et al. How big is the potato (*Solanum tuberosum* L.) yield gap in Sub-Saharan Africa and why? A participatory approach. *Open Agric.* **2018**, *3*, 180–189. [CrossRef]

46. Thomas-Sharma, S.; Abdurahman, A.; Ali, S.; Andrade-Piedra, J.L.; Bao, S.; Charkowski, A.O.; Crook, D.; Kadian, M.; Kromann, P.; Struik, P.C.; et al. Seed degeneration in potato: The need for an integrated seed health strategy to mitigate the problem in developing countries. *Plant Pathol.* **2016**, *65*, 3–16. [CrossRef]

47. Were, H.K.; Kabira, J.N.; Kinyua, Z.M.; Olubayo, F.M.; Karinga, J.K.; Aura, J.; Lees, A.K.; Cowan, G.; Torrance, L. Occurrence and distribution of potato pests and diseases in Kenya. *Potato Res.* **2013**, *56*, 325–342. [CrossRef]

48. Schulte-Geldermann, E.; Gildemacher, P.R.; Struik, P.C. Improving Seed Health and Seed Performance by Positive Selection in Three Kenyan Potato Varieties. *Am. J. Potato Res.* **2012**, *89*, 429–437. [CrossRef]

49. Gebhardt, C.; Valkonen, J.P.T. Organization of genes controlling disease resistance in the potato genome. *Annu. Rev. Phytopathol.* **2001**, *39*, 79–102. [CrossRef]

50. Torrance, L.; Cowan, G.H.; McLean, K.; MacFarlane, S.; Al-Abedy, A.N.; Armstrong, M.; Lim, T.Y.; Hein, I.; Bryan, G.J. Natural resistance to Potato virus Y in Solanum tuberosum Group Phureja. *Theor. Appl. Genet.* **2020**, *133*, 967–980. [CrossRef]

51. Baebler, Š.; Coll, A.; Gruden, K. Plant molecular responses to Potato Virus Y: A continuum of outcomes from sensitivity and tolerance to resistance. *Viruses* **2020**, *12*, 217. [CrossRef]

52. Niehl, A.; Heinlein, M. Perception of double-stranded RNA in plant antiviral immunity. *Mol. Plant Pathol.* **2019**, *20*, 1203–1210. [CrossRef]

53. Nicaise, V. Crop immunity against viruses: Outcomes and future challenges. *Front. Plant Sci.* **2014**, *5*, 660. [CrossRef]

54. Grech-Baran, M.; Witek, K.; Szajko, K.; Witek, A.I.; Morgiewicz, K.; Wasilewicz-Flis, I.; Jakuczun, H.; Marczewski, W.; Jones, J.D.G.; Hennig, J. Extreme resistance to *Potato virus Y* in potato carrying the *Ry$_{sto}$* gene is mediated by a TIR-NLR immune receptor. *Plant Biotechnol. J.* **2020**, *18*, 655–667. [CrossRef]

55. Baebler, Š.; Stare, K.; Kovač, M.; Blejec, A.; Prezelj, N.; Stare, T.; Kogovšek, P.; Pompe-Novak, M.; Rosahl, S.; Ravnikar, M.; et al. Dynamics of responses in compatible potato-Potato virus Y interaction are modulated by salicylic acid. *PLoS ONE* **2011**, *6*, e29009. [CrossRef]

56. Kogovšek, P.; Ravnikar, M. Physiology of the potato–Potato virus Y interaction. *Prog. Bot.* **2013**, 101–133. [CrossRef]

57. Carr, J.P.; Murphy, A.M.; Tungadi, T.; Yoon, J.-Y. Plant defense signals: Players and pawns in plant-virus-vector interactions. *Plant Sci.* **2019**, *279*, 87–95. [CrossRef]

58. Makarova, S.; Makhotenko, A.; Spechenkova, N.; Love, A.J.; Kalinina, N.O.; Taliansky, M. Interactive responses of potato (*Solanum tuberosum* L.) plants to heat stress and infection with *Potato Virus Y*. *Front. Microbiol.* **2018**, *9*, 2582. [CrossRef]

59. Baulcombe, D. RNA silencing. *Trends Biochem. Sci.* **2005**, *30*, 290–293. [CrossRef]

60. Ding, S.W. RNA-based antiviral immunity. *Nat. Rev. Immunol.* **2010**, *10*, 632–644. [CrossRef]

61. Guo, Z.; Li, Y.; Ding, S.W. Small RNA-based antimicrobial immunity. *Nat. Rev. Immunol.* **2019**, *19*, 31–44. [CrossRef] [PubMed]

62. Anandalakshmi, R.; Pruss, G.J.; Ge, X.; Marathe, R.; Mallory, A.C.; Smith, T.H.; Vance, V.B. A viral suppressor of gene silencing in plants. *Proc. Natl. Acad. Sci. USA* **1998**, *95*, 13079–13084. [CrossRef] [PubMed]

63. Mäkinen, K.; De, S. The significance of methionine cycle enzymes in plant virus infections. *Curr. Opin. Plant Biol.* **2019**, *50*, 67–75.

64. Fesenko, I.; Spechenkova, N.; Mamaeva, A.; Makhotenko, A.V.; Love, A.J.; Kalinina, N.O.; Taliansky, M. Role of the methionine cycle in the temperature-sensitive responses of potato plants to potato virus Y. *Mol. Plant Pathol.* **2020**. [CrossRef] [PubMed]

65. Li, J.; Yang, Z.; Yu, B.; Liu, J.; Chen, X. Methylation protects miRNAs and siRNAs from a 3′-end uridylation activity in Arabidopsis. *Curr. Biol.* **2005**, *15*, 1501–1507. [CrossRef]

66. Wang, C.; Wang, C.; Zou, J.; Yang, Y.; Li, Z.; Zhu, S. Epigenetics in the plant–virus interaction. *Plant Cell Rep.* **2019**, *38*, 1031–1038. [CrossRef]

67. Corrêa, R.L.; Sanz-Carbonell, A.; Kogej, Z.; Müller, S.Y.; Ambrós, S.; López-Gomollón, S.; Gómez, G.; Baulcombe, D.C.; Elena, S.F. Viral fitness determines the magnitude of transcriptomic and epigenomic reprogramming of defense responses in plants. *Mol. Biol. Evol.* **2020**, *37*, 1866–1881. [CrossRef]

68. Müller, M.; Munné-Bosch, S. Ethylene response factors: A key regulatory hub in hormone and stress signaling. *Plant Physiol.* **2015**, *169*, 32–41. [CrossRef]

69. Ivanov, K.I.; Eskelin, K.; Bašić, M.; De, S.; Lõhmus, A.; Varjosalo, M.; Mäkinen, K. Molecular insights into the function of the viral RNA silencing suppressor HCPro. *Plant J.* **2016**, *85*, 30–45. [CrossRef]

70. Ruffel, S.; Dussault, M.H.; Palloix, A.; Moury, B.; Bendahmane, A.; Robaglia, C.; Caranta, C. A natural recessive resistance gene against potato virus Y in pepper corresponds to the eukaryotic initiation factor 4E (eIF4E). *Plant J.* **2002**, *32*, 1067–1075. [CrossRef]

71. Parrella, G.; Ruffel, S.; Moretti, A.; Morel, C.; Palloix, A.; Caranta, C. Recessive resistance genes against potyviruses are localized in colinear genomic regions of the tomato (*Lycopersicon* spp.) and pepper (*Capsicum* spp.) genomes. *Theor. Appl. Genet.* **2002**, *105*, 855–861. [CrossRef] [PubMed]

72. Sanchez, P.A.; Babujee, L.; Mesa, H.J.; Arcibal, E.; Gannon, M.; Halterman, D.; Jahn, M.; Jiang, J.; Rakotondrafara, A.M. Overexpression of a modified eIF4E regulates potato virus Y resistance at the transcriptional level in potato. *BMC Genom.* **2020**, *21*, 18. [CrossRef]

73. Shaw, J.; Love, A.J.; Makarova, S.S.; Kalinina, N.O.; Harrison, B.D.; Taliansky, M.E. Coilin, the signature protein of Cajal bodies, differentially modulates the interactions of plants with viruses in widely different taxa. *Nucleus* **2014**, *5*, 85–94. [CrossRef] [PubMed]

74. Lee, W.S.; Fu, S.F.; Li, Z.; Murphy, A.M.; Dobson, E.A.; Garland, L.; Chaluvadi, S.R.; Lewsey, M.G.; Nelson, R.S.; Carr, J.P. Salicylic acid treatment and expression of an *RNA-dependent RNA polymerase* 1 transgene inhibit lethal symptoms and meristem invasion during Tobacco mosaic virus infection in *Nicotiana benthamiana*. *BMC Plant Biol.* **2016**, *16*, 15. [CrossRef] [PubMed]

75. Alamillo, J.M.; Saénz, P.; García, J.A. Salicylic acid-mediated and RNA-silencing defense mechanisms cooperate in the restriction of systemic spread of plum pox virus in tobacco. *Plant J.* **2006**, *2*, 217–227. [CrossRef] [PubMed]

76. Jovel, J.; Walker, M.; Sanfaçon, H. Salicylic acid-dependent restriction of *Tomato ringspot virus* spread in tobacco is accompanied by a hypersensitive response, local RNA silencing, and moderate systemic resistance. *Mol. Plant Microbe Interact.* **2011**, *6*, 706–718. [CrossRef] [PubMed]

77. González, I.; Rakitina, D.; Semashko, M.; Taliansky, M.; Praveen, S.; Palukaitis, P.; Carr, J.P.; Kalinina, N.; Canto, T. RNA binding is more critical to the suppression of silencing function of Cucumber mosaic virus 2b protein than nuclear localization. *RNA* **2012**, *18*, 771–782. [CrossRef]

78. Van Dam, J.; Kooman, P.L.; Struik, P.C. Effects of temperature and photoperiod on early growth and final number of tubers in potato (*Solanum tuberosum* L.). *Potato Res.* **1996**, *39*, 51–62. [CrossRef]

79. Li, J.; Lin, X.; Chen, A.; Peterson, T.; Ma, K.; Bertzky, M.; Ciasis, P.; Kapos, V.; Peng, C.; Poulter, B. Global priority conservation areas in the face of 21st century climate change. *PLoS ONE* **2013**, *8*, e54839. [CrossRef]

80. Trapero-Mozos, A.; Morris, W.L.; Ducreux, L.J.; McLean, K.; Stephens, J.; Torrance, L.; Bryan, G.J.; Hancock, R.D.; Taylor, M.A. Engineering heat tolerance in potato by temperature dependent expression of a specific allele of heat-shock cognate 70. *Plant Biotechnol. J.* **2018**, *16*, 197–207. [CrossRef]

81. Prasch, C.M.; Sonnewald, U. Simultaneous application of heat, drought, and virus to Arabidopsis plants reveals significant shifts in signaling networks. *Plant Physiol.* **2013**, *162*, 1849–1866. [CrossRef] [PubMed]

82. Bradshaw, J.E.; Ramsay, G. Utilisation of the Commonwealth Potato Collection in potato breeding. *Euphytica* **2005**, *146*, 9–19. [CrossRef]

83. Solomon-Blackburn, R.M.; Bradshaw, J.E. Resistance to Potato virus Y in a multi-trait potato breeding scheme without direct selection in each generation. *Potato Res.* **2007**, *50*, 87–95. [CrossRef]

84. Zhu, Y.; Qian, W.; Hua, J. Temperature modulates plant defense responses through NB-LRR proteins. *PLoS Pathog.* **2010**, *6*, e1000844. [CrossRef]

85. Szittya, G.; Silhavy, D.; Molnár, A.; Havelda, Z.; Lovas, Á.; Lakatos, L.; Bánfalvi, Z.; Burgyán, J. Low temperature inhibits RNA silencing mediated defence by the control of siRNA generation. *EMBO J.* **2003**, *22*, 633–640. [CrossRef]

86. Shams-Bakhsh, M.; Canto, T.; Palukaitis, P. Enhanced resistance and neutralization of defense responses by suppressors of RNA silencing. *Virus Res.* **2007**, *130*, 103–109. [CrossRef]

87. Nazar, R.; Iqbal, N.; Umar, S. Heat Stress Tolerance in Plants: Action of Salicylic Acid pp145-161. In *Salicylic Acid: A Multifaceted Hormone*; Nazar, R., Iqbal, N., Khan, N., Eds.; Springer: Singapore, 2017.

88. Hofius, D.; Maier, A.T.; Dietrich, C.; Jungkunz, I.; Börnke, F.; Maiss, E.; Sonnewald, U. Capsid protein-mediated recruitment of host DnaJ-like proteins is required for Potato virus Y infection in tobacco plants. *J. Virol.* **2007**, *81*, 11870–11880. [CrossRef]

89. Mäkinen, K.; Hafrén, A. Intracellular coordination of potyviral RNA functions in infection. *Front. Plant Sci.* **2014**, *5*, 110. [CrossRef]

90. Whitham, S.A.; Quan, S.; Chang, H.S.; Cooper, B.; Estes, B.; Zhu, T.; Wang, X.; Hou, Y.M. Diverse RNA viruses elicit the expression of common sets of genes in susceptible Arabidopsis thaliana plants. *Plant J.* **2003**, *33*, 271–283. [CrossRef]

91. Bonierbale, M.W.; Amoros, W.R.; Salas, E.; de Jong, W. Potato Breeding. In *The Potato Crop: Its Agricultural, Nutritional and Social Contribution to Humankind*; Campos, H., Ortiz, O., Eds.; Springer: Berlin/Heidelberg, Germany, 2020.

92. Zimnoch-Guzowska, E.; Yin, Z.; Chrzanowska, M.; Flis, B. Sources and Effectiveness of Potato PVY Resistance in IHAR's Breeding Research. *Am. J. Potato Res.* **2013**, *90*, 21–27. [CrossRef]

93. Valkonen, J.P.T.; Gebhardt, C.; Zimnoch-Guzowska, E.; Watanabe, K.N. Resistance to Potato virus Y in Potato. In *Potato virus Y: Biodiversity, Pathogenicity, Epidemiology and Management*; Lacomme, C., Glais, L., Bellstedt, D., Dupuis, B., Karasev, A., Jacquot, E., Eds.; Springer: Berlin/Heidelberg, Germany, 2017.

94. Koonin, E.V.; Makarova, K.S.; Zhang, F. Diversity, classification and evolution of CRISPR-Cas systems. *Curr. Opin. Microbiol.* **2017**, *37*, 7–78. [CrossRef] [PubMed]

95. Kalinina, N.O.; Khromov, A.; Love, A.J.; Taliansky, M.E. CRISPR Applications in plant virology: Virus resistance and beyond. *Phytopathology* **2020**, *110*, 18–28. [CrossRef] [PubMed]

96. Zhan, X.; Zhang, F.; Zhong, Z.; Chen, R.; Wang, Y.; Chang, L.; Bock, R.; Nie, B.; Zhang, J. Generation of virus resistant potato plants by RNA genome targeting. *Plant Biotechnol. J.* **2019**, *17*, 1814–1822. [CrossRef] [PubMed]

97. Makhotenko, A.V.; Khromov, A.V.; Snigir, E.A.; Makarova, S.S.; Makarov, V.V.; Suprunova, T.P.; Kalinina, N.O.; Taliansky, M.E. Functional analysis of coilin in virus resistance and stress tolerance of potato *Solanum tuberosum* using CRISPR-Cas9 editing. *Dokl. Biochem. Biophys.* **2019**, *484*, 88–91. [CrossRef] [PubMed]

Article

Transmission of the Bean-Associated Cytorhabdovirus by the Whitefly *Bemisia tabaci* MEAM1

Bruna Pinheiro-Lima [1,2,3], Rita C. Pereira-Carvalho [2], Dione M. T. Alves-Freitas [1], Elliot W. Kitajima [4], Andreza H. Vidal [1,3], Cristiano Lacorte [1], Marcio T. Godinho [1], Rafaela S. Fontenele [5], Josias C. Faria [6], Emanuel F. M. Abreu [1], Arvind Varsani [5,7], Simone G. Ribeiro [1,*] and Fernando L. Melo [2,3,*]

[1] Embrapa Recursos Genéticos e Biotecnologia, Brasília DF 70770-017, Brazil; pinheiro.limab@gmail.com (B.P.-L.); dionebio@gmail.com (D.M.T.A.-F.); andrezactg@hotmail.com (A.H.V.); cristiano.lacorte@embrapa.br (C.L.); marcioslash@gmail.com (M.T.G.); emanuel.abreu@embrapa.br (E.F.M.A.)

[2] Departamento de Fitopatologia, Instituto de Biologia, Universidade de Brasília, Brasília DF 70275-970, Brazil; rcpcarvalho@unb.br

[3] Departamento de Biologia Celular, Instituto de Biologia, Universidade de Brasília, Brasília DF 70275-970, Brazil

[4] Departamento de Fitopatologia, Escola Superior de Agricultura Luiz de Queiroz, Piracicaba SP 13418-900, Brazil; ewkitaji@usp.br

[5] The Biodesign Center for Fundamental and Applied Microbiomics, Center for Evolution and Medicine School of Life Sciences, Arizona State University, Tempe, AZ 85287-5001, USA; rafasfontenele@gmail.com (R.S.F.); avarsani@gmail.com (A.V.)

[6] Embrapa Arroz e Feijão, Goiânia GO 75375-000, Brazil; josias.faria@embrapa.br

[7] Structural Biology Research Unit, Department of Integrative Biomedical Sciences, University of Cape Town, Observatory, Cape Town 7701, South Africa

* Correspondence: simone.ribeiro@embrapa.br (S.G.R.); flucasmelo@gmail.com (F.L.M.)

Received: 4 August 2020; Accepted: 11 September 2020; Published: 15 September 2020

Abstract: The knowledge of genomic data of new plant viruses is increasing exponentially; however, some aspects of their biology, such as vectors and host range, remain mostly unknown. This information is crucial for the understanding of virus–plant interactions, control strategies, and mechanisms to prevent outbreaks. Typically, rhabdoviruses infect monocot and dicot plants and are vectored in nature by hemipteran sap-sucking insects, including aphids, leafhoppers, and planthoppers. However, several strains of a potentially whitefly-transmitted virus, papaya cytorhabdovirus, were recently described: (i) bean-associated cytorhabdovirus (BaCV) in Brazil, (ii) papaya virus E (PpVE) in Ecuador, and (iii) citrus-associated rhabdovirus (CiaRV) in China. Here, we examine the potential of the *Bemisia tabaci* Middle East-Asia Minor 1 (MEAM1) to transmit BaCV, its morphological and cytopathological characteristics, and assess the incidence of BaCV across bean producing areas in Brazil. Our results show that BaCV is efficiently transmitted, in experimental conditions, by *B. tabaci* MEAM1 to bean cultivars, and with lower efficiency to cowpea and soybean. Moreover, we detected BaCV RNA in viruliferous whiteflies but we were unable to visualize viral particles or viroplasm in the whitefly tissues. BaCV could not be singly isolated for pathogenicity tests, identification of the induced symptoms, and the transmission assay. BaCV was detected in five out of the seven states in Brazil included in our study, suggesting that it is widely distributed throughout bean producing areas in the country. This is the first report of a whitefly-transmitted rhabdovirus.

Keywords: common bean; *Phaseolus vulgaris*; cytorhabdovirus; whitefly; *Bemisia tabaci*; vector; virus transmission; virus evolution

---

1. Introduction

Rhabdoviruses (family *Rhabdoviridae*) are a group of negative-sense, single-stranded RNA viruses that infect plants, vertebrate animals, and invertebrate animals. They cause harmful diseases in humans and animals and can cause high yield losses in crops. Plant-infecting rhabdoviruses are currently taxonomically assigned to six genera [1]. Members of the *Dichorhavirus* genus have a bi-segmented genome, infect di- and monocotyledonous plants, and are transmitted by *Brevipalpus* mites. Viruses belonging to the *Varicosavirus* genus also have a bi-segmented genome, infect plants of the Compositae and Solanaceae families, and are transmitted by zoospores of the fungus *Olpidium brassicae*. Non-segmented plant rhabdoviruses infect mono- and dicot plants and are vectored in nature by hemipteran sap-sucking insects, including aphids, leafhoppers, and planthoppers [2]. Moreover, there is a close relationship between plant rhabdoviruses and their vectors, and each virus may be vectored by one species or a few closely related ones [3]. Viruses that replicate within the nuclei of infected plant cells are assigned to the genera *Alphanucleorhabdovirus*, *Betanucleorhabdovirus*, and *Gammanucleorhabdovirus*, while those that multiply in the cell cytoplasm belong to the *Cytorhabdovirus* [2,4]. Insect and mite-transmitted rhabdoviruses also replicate in their arthropod vectors and are transmitted in a persistent propagative manner [3,5]. However, no information on vectors and transmission characteristics is available for most of these viruses [5].

Bean-associated cytorhabdovirus (BaCV) was identified in transgenic bean golden mosaic virus (BGMV)-resistant common bean lines [6]. The BaCV genome has a 3′-N-P-M-G-L-5′ [nucleocapsidprotein (N), phosphoprotein (P), matrix protein (M), glycoprotein (G), and RNA-dependent RNA polymerase (RdRp) protein (L)] organization that is typical of rhabdoviruses and between P and M, the BaCV genome encodes two accessory genes, P3 and P4 [6]. A closely related virus with a genome sequence identity of 97% named papaya virus E (PpVE) has been reported from papaya plants in Ecuador [7]. Based on the high sequence identity between the two virus sequences, it was proposed that the virus species would be named *Papaya cytorhabdovirus* with strains PpVE infecting papayas and BaCV infecting beans [8,9].

To characterize the bean-infecting cytorhabdovirus strain in detail, we carried out the molecular cloning and defined its morphological and cytopathological characteristics. Moreover, we observed a high prevalence of BaCV in bean fields in Brazil and determined the efficient transmission of BaCV by the whitefly *Bemisia tabaci* Middle East-Asia Minor 1 (MEAM1), a hitherto unknown feature for a plant rhabdovirus.

2. Materials and Methods

2.1. Plant Material

Fifteen common bean plants (cultivar 'Pérola') with typical virus symptoms of mosaic, leaf distortion, crumpling, and dwarfing (Figure 1) were collected in a commercial field in Luziânia, Goiás State, in June 2016. Leaf samples from the 15 plants were detached and stored at −80 °C. Six plants were transplanted to soil-filled pots and maintained in a screen-protected cage, without removing the abundant whitefly colonies present in the plant leaves. One of the transplanted plants survived and was used for virus transmission and cloning of the genome of BaCV-Luz.

Figure 1. Symptoms in common bean plants collected in a commercial field in Luziânia, Goiás state with mixed infection by bean-associated cytorhabdovirus (BaCV), cowpea mild mottle virus (CPMMV), and bean golden mosaic virus (BGMV) and whitefly colonization. (**a**) Mosaic and leaf wrinkling. (**b**) Mosaic, severe leaf distortion, and deformation. (**c**) Detail of leaves with reduced area, yellow mosaic, and severe crinkling and curling.

2.2. Distribution of BaCV in Common Beans in Brazil

Initially, the 15 plants collected in Luziânia were examined by RT-PCR or PCR for the presence of bean-infecting viruses frequently found in Brazil, the whitefly-transmitted CPMMV (family *Betaflexiviridae*; genus *Carlavirus*), BGMV, and macroptilium yellow spot virus—MaYSV (family *Geminiviridae*; genus *Begomovirus*), the chrysomelid beetle-transmitted bean rugose mosaic virus—BRMV (family *Secoviridae*; genus *Comovirus*), and the recently identified cytorhabdovirus BaCV. To determine the occurrence of BaCV in different areas in Brazil (Figure 2), additional bean samples were collected from experimental or commercial bean fields between 2016 to 2018, including Brasília ($n = 30$), Distrito Federal—DF; Santo Antônio de Goiás ($n = 26$), Luziânia ($n = 15$), Cristalina ($n = 43$), Urutai ($n = 1$), and Araçu ($n = 1$) in Goiás State—GO; Sorriso, Mato Grosso State—MT ($n = 2$); Bonfinópolis de Minas ($n = 3$), Paracatu ($n = 1$), Três Pontas ($n = 5$) in Minas Gerais State; and Arapiraca ($n = 5$) in Alagoas State. In addition, we analyzed bean samples from our archived collection. Plants collected in Brasília, DF in 2007 ($n = 2$), PAD/DF Paranoá, DF in 2012 ($n = 11$), and Riacho Fundo, DF in 2015 ($n = 41$); Cruz das Almas ($n = 8$), Morro do Chapéu ($n = 4$), Piritiba ($n = 3$), and Antônio Gonçalves ($n = 2$) in Bahia State in 2015; and Palmital ($n = 1$) in São Paulo State in 2015 were screened for BaCV infection. In total, 219 bean plants were screened for the presence of BaCV, CPMMV, and BGMV (41 BGMV-immune cultivar 'BRS FC 401 RMD' plants were not tested for BGMV).

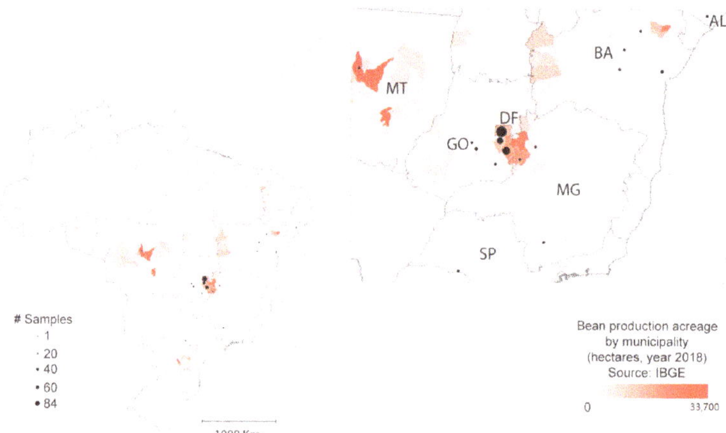

Figure 2. Summary of the distribution of the bean sampling locations at Alagoas (AL), Bahia (BA), Goiás (GO), Minas Gerais (MG), Mato Grosso (MT), São Paulo (SP) states and Distrito Federal (DF). Each municipality was colored in red according to the bean planted area (in hectare), obtained from the Brazilian Institute of Geography (https://www.ibge.gov.br/) [10]. The black circles represent the sample sites, and their size is proportional to the number of collected plants. The samples were collected between 2007 and 2018.

2.3. RNA and DNA Extraction

Total RNA was extracted from ~100 mg of plant leaf tissue (pulverized in liquid nitrogen) using the TRIzol Reagent (Invitrogen, Carlsbad, CA, USA) according to the manufacturer's instructions. Total RNA was also extracted from (i) a group of 30 whitefly individuals, (ii) from one single individual, and (iii) parts of one single whitefly (head/thorax or abdomen). One fourth and one-tenth of the reagent's volumes of the standard TRIzol Reagent protocol were used for 30 whiteflies and one or parts of an insect, respectively. Total DNA was extracted from leaves and whiteflies using the CTAB protocol [11].

2.4. RT-PCR, PCR, and Cloning

The sequences and characteristics of all primers used in this study are summarized in Table S1. To determine the complete genome sequence of BaCV-Luz, first we determined the 5′ and 3′ ends by rapid amplification of cDNA ends (RACE) as previously described [6,12,13]. Next, based on the genome sequence of BaCV-GO [6] and BaCV-Luz leader and trailer sequences, primers were designed to amplify the virus genome by RT-PCR in six fragments overlapping in at least 150 nt. The cDNA was synthesized with 5 µL of total RNA (approximately 1 µg) using SuperScript III Reverse Transcriptase (Invitrogen, Carlsbad, CA, USA), Anchored Oligo(dT)20 (Invitrogen, Carlsbad, CA, USA) and random primers [14]. One microliter of cDNA was used in PCR reactions with LongAmp Taq DNA Polymerase (New England Biolabs, Ipswich, MA, USA) and specific primers for each fragment (Table S1). Amplicons were gel-purified, cloned into PCR 2.1 TOPO TA vector (Invitrogen, Carlsbad, CA, USA) and sequenced. At least two clones of each fragment were sequenced at Macrogen Inc. (Seoul, Korea).

Detection of BaCV (in plants and whiteflies), CPMMV (in plants and whiteflies), and BRMV (in plants) was carried out by RT-PCR using the cDNA prepared as described above as template. PCR reactions were performed with Taq DNA Polymerase (Invitrogen, Carlsbad, CA, USA) using 1 µL of cDNA and primers BaCV_1F/BaCV_1579R, CPMMV_4000F/CPMMV_4500R [15], and BRMV1_76F/BRMV1_521R [6] specific for BaCV, CPMMV, and BRMV, respectively. For BGMV and MaYSV PCR-based infection surveys of field samples, total extracted plant DNA was used as template and primers BGMV_HPXHO/BGMV_HPKPN for BGMV [16] and MaYSV-249F/MaYSV-1083R for MaYSV. A portion of the PCR amplicons was verified by cloning in PCR 2.1 TOPO TA vector (Invitrogen, Carlsbad, CA, USA) and Sanger sequencing.

Whiteflies populations from the field, as well as from the Universidade de Brasília rearing facility, were identified as *B. tabaci* MEAM1 by genotyping the insects using PCR-RFLP as described by Bosco, et al. [17]. Amplification of a region of the mitochondrial cytochrome oxidase I gene (mtCOI) was undertaken by PCR using whitefly total DNA as template and the primers COI-Fw/COI-Rv [17] (Table S1). The amplicons were digested with Taq I endonuclease (New England Biolabs, Ipswich, MA, USA). The digestion products were resolved in a 1.2% agarose gel electrophoresis, and the profile of the bands was compared with the profiles for *B. tabaci* MEAM1 (synonym biotype B) and *B. tabaci* MED (synonym biotype Q) [17].

2.5. Sequence Analysis

All sequences generated in this study were trimmed and assembled using Geneious software (v. 11, Biomatters, Auckland, New Zealand) [18]. The sequence identity was confirmed by BLASTn analysis [19]. The complete genome of BaCV-Luz isolate was deposited in GenBank under the accession number MT811775. The RdRp amino acid sequence of BaCV-Luz was aligned with those of all cytorhabdoviruses available in GenBank (as by May 2020) (Table S2) using MAFFT algorithm [20] implemented in Geneious. Maximum likelihood (ML) phylogenetic tree was inferred using IQ-TREE [21], with node support estimated with the Shimodaira–Hasegawa approximate likelihood ratio test (SH-aLRT) [22]. Moreover, the amino acid sequences of the glycoproteins encoded by the cytorhabdoviruses were used to generate a sequence similarity network using the Enzyme Function Initiative–Enzyme Similarity Tool (EFI–EST) [23] with an alignment score threshold of 35 and minimum E-value threshold of 1×10^{-5}. The network was visualized in Cytoscape v3.7.1 [24]. Pairwise sequence identity comparisons were performed using the Sequence Demarcation Tool (SDT) v.1.2 [25].

2.6. BaCV Transmission by B. tabaci MEAM1

After confirming the infection by BaCV, CPMMV, and BGMV, a bean plant colonized by whiteflies transplanted from the field was used as an inoculum source for transmission tests (Figure 3). Initially, eight young bean seedlings each of cultivars 'Pérola', 'Jalo', and 'BRS FC 401 RMD' were placed in the cage with the whitefly-infested plant. Seven days later, another set of eight seedlings of each cultivar was placed inside the same cage. Plant samples were collected 14 days after their introduction into the cage and tested for the presence of BaCV, CPMMV, and BGMV. Since 'BRS FC 401 RMD' plants are immune to BGMV [26], they were not tested for BGMV infection. Whiteflies were also collected and tested for the presence of BaCV. Following this, the BaCV-Luz isolate was maintained in 'BRS FC 401 RMD' plants by introducing plantlets every three to four weeks in the cage, and when the population of whiteflies was declining, adult individuals from the rearing facility from the Universidade de Brasília were added to the cage for BaCV-Luz isolate maintenance until 2018.

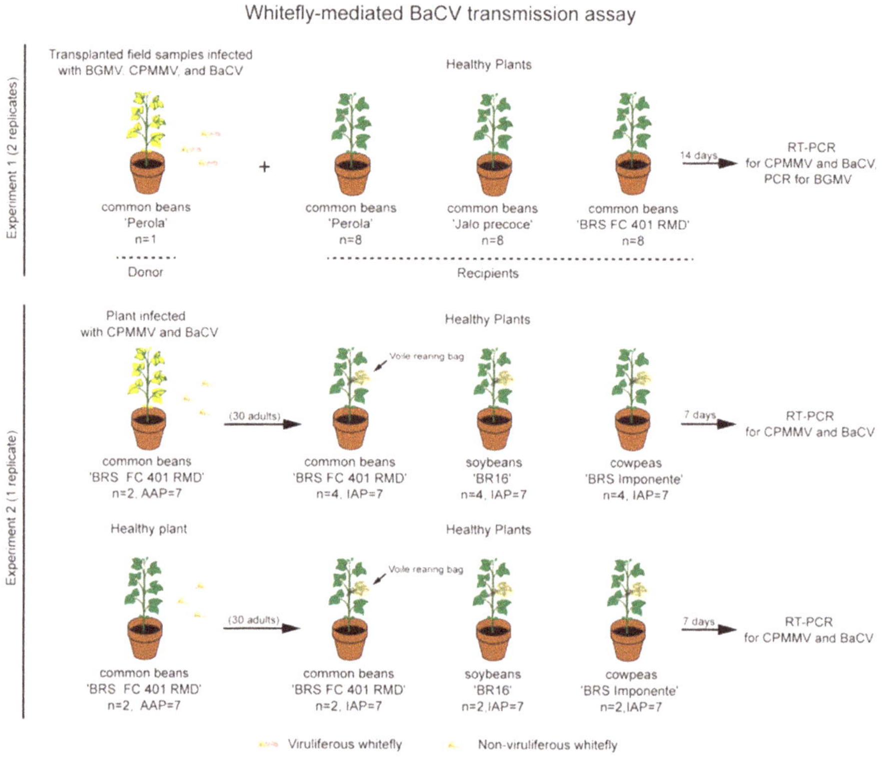

Figure 3. Whitefly (*Bemisia tabaci* Middle East-Asia Minor 1 (MEAM1)) mediated transmission of BaCV. Outline of experimental procedures: In experiment 1, a common bean plant with whitefly colonies from the field was used as an inoculum source for BaCV (BGMV and CPMMV) transmission to bean plants 'Jalo', 'Pérola' and 'BRS FC 401 RMD'. After 14 days, samples were collected, and BaCV (CPMMV and BGMV) transmission confirmed by RT-PCR and PCR. In experiment 2, whiteflies were exposed to BaCV- (and CPMMV)-infected plants for a 7-days acquisition period. Thirty whiteflies were transferred from the inoculum source plant to a healthy plant leaf wrapped by a voile rearing bag for a 7-days inoculation period. After seven days, samples were collected and tested by RT-PCR. The same procedure was applied to healthy plants with aviruliferous whiteflies as controls.

To further refine the transmission tests, two three-week old 'BRS FC 401 RMD' plants infected with BaCV-Luz (and CPMMV) were removed from the cage, sprayed with imidacloprid (32 mg/L) to eliminate any whiteflies present in the plants, transferred to a bugdorm tent, and kept in a greenhouse at room temperature for 10 days. *B. tabaci* MEAM1 individuals were reared in cabbage plants (*Brassica oleraceae* var. *capitata* L.) at the Biological Experimental Station of the Universidade de Brasília. To synchronize the whiteflies' age, adult flies were removed from the cabbage plants, the plants were placed in a bugdorm cage, and kept in a greenhouse for three days. One-to-three-day old adult whiteflies were collected from the cabbage plants and transferred to the BaCV-Luz-infected bean plants for an acquisition accession period (AAP) of 7 days (Figure 3). As a control, a batch of whiteflies was placed for 7 days in healthy bean plants. Next, 25 to 30 potentially viruliferous whiteflies were transferred from the virus-infected source plants to a polyester voile rearing bag previously placed around a trifoliate leaf of 4 healthy beans 'BRS FC 401 RMD', 4 healthy soybeans [*Glycine max* (L.) Merr. 'BR16'], and 4 healthy cowpeas [*Vigna unguiculata* (L.) Walp. 'BRS Imponente'] plants for an inoculation access period (IAP) of 7 days. Likewise, as controls, non-viruliferous flies were placed in 2 healthy plants of each plant species for 7 days. After this period, the leaves inside the polyester voile bags were detached from plants to avoid nymphal maturation, and plants were sprayed with imidacloprid (32 mg/L) to remove any possible remaining whiteflies [27]. Plants were observed for the development of symptoms, and virus infection was confirmed by RT-PCR, as described above.

2.7. BaCV Detection in Plants and Whiteflies

To identify the presence of BaCV-Luz in bean plants and whiteflies, the total RNA was treated with TURBO DNase (Invitrogen, Carlsbad, CA, USA) to eliminate any DNA trace from the RNA preparation as described by Cao, et al. [28]. The cDNA was prepared with Anchored Oligo (dT) 20 primer (Invitrogen, Carlsbad, CA, USA) and SuperScript III Reverse Transcriptase (Invitrogen, Carlsbad, CA, USA) and the PCR reactions were performed with primers specific for all BaCV genes (Table S1). The transcripts for Actin-11 (*act11*) [29] and the small Rubisco subunit (*RbcS*) [30] genes from common bean, and Ribosomal protein L9 (*RpL9*) [31] and Vacuolar ATPase (*v-ATPase*) subunit A [31] genes from whitefly, were used as internal reference controls and to identify possible transcripts ingested by the whiteflies during the feeding in the bean plants (Table S1).

2.8. Transmission Electron Microscopy

Small leaf pieces were cut from 'BRS FC 401 RMD' bean plants that were experimentally inoculated with whitefly and tested positive for BaCV by RT-PCR. Leaf sections were fixed overnight with Karnovsky modified fixative (2.5% glutaraldehyde and 2% paraformaldehyde in 0.05 M cacodylate buffer, pH 7.2). The samples were post-fixed with 1% osmium tetroxide (in 0.05 M cacodylate buffer) for 1–2 h. Tissues were dehydrated, embedded in low viscosity epoxy Spurr resin (Electron Microscopy Sciences, Hatfield, PA, USA), and sectioned in a Leica UC6 ultramicrotome equipped with a Diatome diamond knife. Ultrathin sections (70–100 nm thick) were transferred onto 300 mesh copper grids, stained with 3% uranyl acetate and Reynold's lead citrate, and examined in a JEOL JEM 1011 transmission electron microscope. Healthy bean leaves were used as controls. Images were digitally recorded. Whiteflies were reared and collected from BaCV-Luz-infected bean plants, dipped into a NaCl 0.9% solution, and dissected at the thorax region in two parts: head/thorax and abdomen. For each whitefly, either the head/thorax or the abdomen parts were used for BaCV detection by RT-PCR, and the other part was fixed in a cold solution of 2.5% glutaraldehyde and 1.8% sucrose in 0.1 M cacodylate buffer. For the whiteflies that one part of the body was positive for BaCV presence by RT-PCR, the other part was processed to be examined by electron microscopy, as described above for plant tissues. Three pairs of head and abdominal parts of presumably viruliferous whiteflies were examined.

3. Results and Discussion

3.1. BaCV Is Widely Distributed throughout Common Bean Producing Areas in Brazil

BaCV was identified for the first time in Brazil in 2014 in bean plants collected in Santo Antônio de Goiás, Goiás State (GO) [32]. In June 2016, severe virus-like symptoms were recorded by farmers in Luziânia (GO) (Figure 1). The incidence of these symptoms in the bean plants ranged from 20% to 80%, depending on the area. The bean fields were heavily infested by whiteflies migrating from nearby cotton and tomato fields. Therefore, the first 15 bean plants received from the farmers were screened by PCR or RT-PCR for the presence of the whitefly-borne viruses BGMV, MaYSV, and CPMMV in addition to BRMV, and the new cytorhabdovirus BaCV (Figure S1). None of the plants were infected with MaYSV or BRMV. Conversely, all 15 plants had a mixed infection with BaCV, BGMV, and CPMMV.

Given the high infection rate, we further investigated the occurrence of these viruses in newly collected (2016–2018) and archived samples (2007–2016) from Central, Southeast, and Northeastern Brazil, comprising six states and the federal district (Figure 2). In total, 219 plants were analyzed, and 91 (41.55%) were infected by BaCV (Table 1 and Figure 4). Most of the plants analyzed (46.12%)

were sampled in Goiás, Central Brazil, one of the top bean-producing States in the country, where the incidence of BaCV reached up to 100%, depending on the area (Table 1). Despite the uneven sampling among regions, we were able to detect BaCV in Southeast and Northeast regions, which are more than 2000 km apart, suggesting that BaCV is widely distributed throughout bean producing areas in the country. However, to have a better view of the prevalence of BaCV in the bean crops in Brazil, further sampling should be conducted in additional bean cultivating areas in the Southeastern, Northeastern, and Southern states.

Table 1. Detection of BaCV, CPMMV, and BGMV in common bean samples collected in Brazil by RT-PCR and PCR.

State/Total Number of Samples	City	Sampled Areas	Year	Virus Incidence (Infected/Tested)		
				BaCV	CPMMV	BGMV
Bahia (BA) (*n* = 17)	Cruz das Almas	1	2015	2/8	8/8	4/8
	Morro do Chapéu	1	2015	0/4	3/4	0/4
	Piritiba	1	2015	0/3	3/3	0/3
	Antônio Gonçalves	1	2015	0/2	2/2	0/2
Total in Bahia				2/17	16/17	4/17
Alagoas (AL) (n = 5)	Arapiraca	1	2018	0/5	4/5	1/5
Total in Alagoas				0/5	4/5	1/5
Minas Gerais (MG) (n = 9)	Bonfinópolis de Minas	1	2016	0/3	3/3	0/3
	Paracatu	1	2016	1/1	1/1	1/1
	Três Pontas	1	2018	0/5	1/5	2/5
Total in Minas Gerais				1/9	5/9	3/9
São Paulo (SP) (n = 1)	Palmital	1	2015	1/1	1/1	1/1
Total in São Paulo				1/1	1/1	1/1
Goiás (GO) (n = 101)	Luziânia	1	2016	15/15	15/15	15/15
		2	2016	7/7	7/7	7/7
		3	2016	3/4	4/4	4/4
		4	2016	4/4	4/4	4/4
	Cristalina	1	2016	10/14	14/14	14/14
		2	2016	0/14	12/14	13/14
		3	2016	11/11	10/11	11/11
		4	2016	4/4	4/4	4/4
	Santo Antônio de Goiás	1	2016	0/4	4/4	4/4
		2	2016	2/17	17/17	8/8
		1	2018	0/5	4/5	4/5
	Urutaí	1	2018	0/1	0/1	0/1
	Araçu	1	2018	0/1	1/1	1/1
Total in Goiás				56/101	96/101	89/92
Distrito Federal (DF) (n = 84)	Brasília	1	2017	8/30	19/30	14/30
		1	2007	0/2	0/2	2/2
		1	2012	0/11	1/11	7/11
		1	2015	8/12	6/12	-
		2	2015	15/29	25/29	-
Total in Distrito Federal				31/84	51/84	23/43
Mato Grosso (MT) (n = 2)	Sorriso	1	2018	0/2	1/2	2/2
Total in Mato Grosso				0/2	1/2	2/2
TOTAL				91/219	174/219	123/169

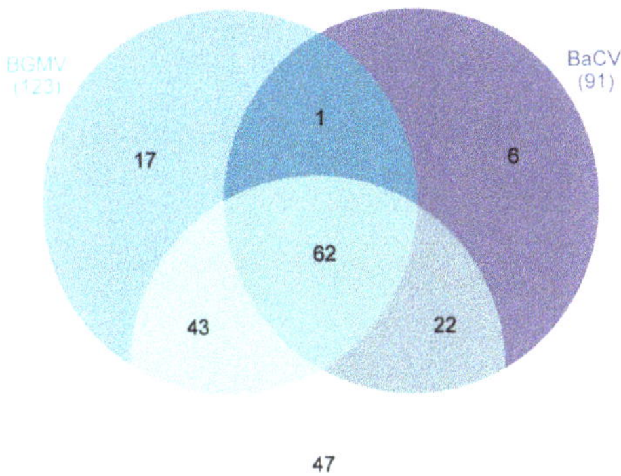

Figure 4. Distribution of viral infections in 219 common bean plants from 2007 to 2018. BaCV was found mostly in mixed infections with BGMV and CPMMV. Only six plants were found singly infected with BaCV.

BaCV was found in a single infection only in six samples of BGMV-resistant 'BRS FC401 RMD' plants collected in Riacho Fundo, DF, in 2015. The remaining 85 BaCV positive plants were co-infected with CPMMV (*n* = 22), BGMV (*n* = 1) or with CPMMV and BGMV (*n* = 62) (Table 1, Figure 4). Mixed infections were, therefore, common in these plants. Viral co-infections are very common and, in the field, seem to be the rule rather than the exception, and may result in synergistic effects and stronger symptoms [33]. Moreover, the change in the plant phenotype induced by the co-infection can alter or increase the attraction of vectors and facilitate the transmission of these viruses and enhance epidemics [34]. Unfortunately, singly BaCV-infected samples were identified only in archived samples, and we do not have records of the specific symptoms displayed by these plants. Therefore, it was impossible to establish possible effects of mixed infections on the severe symptoms observed in the field. Importantly, the phenotype of plants in the field is also influenced by other biotic and abiotic factors such as water and temperature stress and infection by other pathogens like bacteria and fungi.

3.2. Evolutionary Analysis Indicates Whiteflies as the Potential Vector for BaCV

The complete BaCV-Luz genome was determined by PCR of six overlapping fragments and was 13,467 nt in length. As expected, the genome presented the seven ORFs originally described in BaCV-GO (N, P, P3, P4, M, G, and L), flanked by two non-transcribed leader and trailer regions (Figure 5). As shown in Table S3, the complete genome of BaCV-Luz shares 99.8% and 96.3% identity with BaCV-GO and PpVE, respectively. To further investigate the phylogenetic relationship of cytorhabdoviruses, we aligned 28 RdRP amino acid sequences (2374 aa in length, including gaps) from reference cytorhabdoviruses available in Genbank, including two sequences derived from transcriptomes of *B. tabaci* (Table S2). The phylogenetic analysis shows that cytorhabdoviruses cluster in monophyletic groups according to its potential vector: aphid [35–39], planthopper [40–45], leafhopper [46], whitefly, and an undescribed vector (Figure 6). As previously suggested [7,47,48], BaCV isolates and PpVE were closely related to *B. tabaci* TSA 2 (AKC57270.1), confirming that these viruses belong to the species *Papaya cytorhabdovirus*. Moreover, yerba mate chlorosis-associated virus clustered with *B. tabaci* TSA 1 (AKC57275.1), as a sister group of papaya cytorhabdovirus strains. Furthermore, assuming that the viral surface envelope glycoprotein G interacts directly with receptors in the vector cells, a sequence similarity network (EFI–EST webserver) was generated using the glycoprotein amino acid sequences encoded by the 28 cytorhabdoviruses. Interestingly, the analyses show three distinct clusters (aphids, planthopper, and whiteflies) and two singletons (leafhopper and an undescribed vector) with a high degree of interconnectivity.

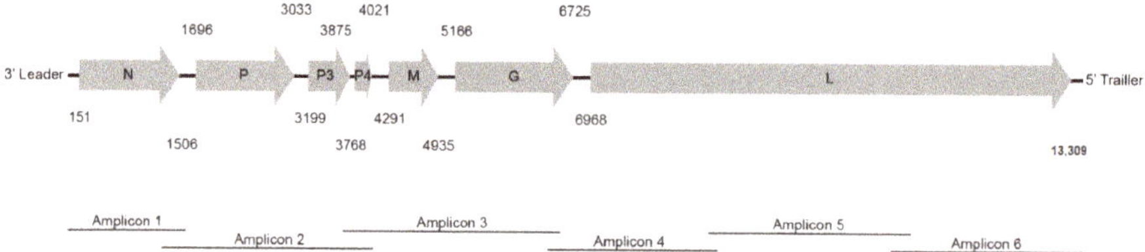

Figure 5. Genome organization of BaCV-Luz. Canonical genes encoding [N] nucleoprotein, [P] phosphoprotein, [M] matrix protein, [G] glycoprotein and [L] RNA-dependent RNA polymerase (RdRp), and non-canonical [P3] and [P4]. Each ORF is represented by a gray arrow with the first and last nucleotide positions depicted. The complete genome (13,467 nts) was recovered by RT-PCR of six overlapping amplicons.

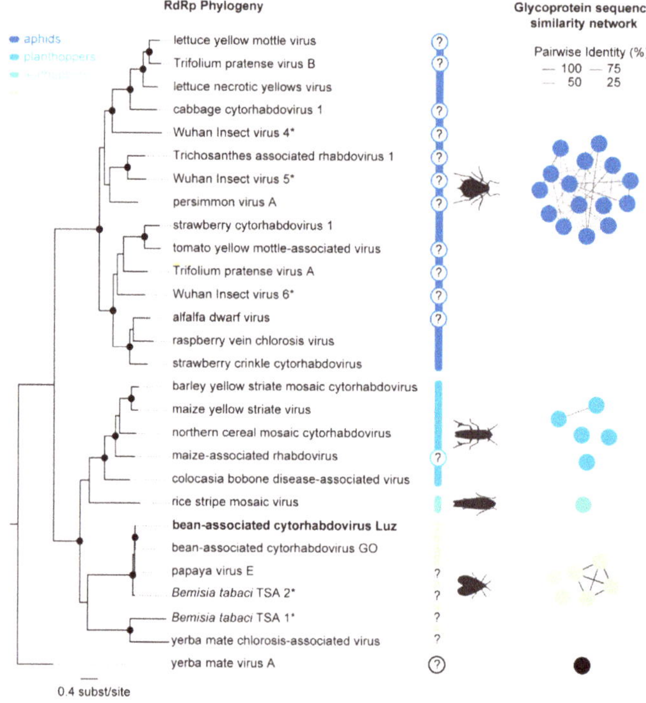

Figure 6. Maximum likelihood phylogenetic tree of cytorhabdovirus RdRP amino acid sequences and sequence similarity network of cytorhabdovirus glycoprotein amino acid sequences encoded by BaCV-Luz, BaCV-GO and other 26 cytorhabdoviruses. In both analyses, four groups were clustered according to their probable vector: aphid, planthopper, leafhopper, and whitefly. Enzyme Function Initiative–Enzyme Similarity Tool (EFI–EST) was used for glycoprotein analysis with an alignment score threshold of 35 and a minimum E-value threshold of 1×10^{-5}. The network was visualized in Cytoscape v3.7.1. Support values $\geq$ 90% SH-aLRT are displayed with black circles at nodes. (?) Unknown vector species. (*) Viruses detected by insects metatranscriptome.

These results, together with the detection of BaCV in concomitant infection with the whitefly vectored BGMV and CPMMV in all plants (Figure 4), and with the presence of whiteflies adults, nymphs, and eggs in these plants (Figure 1), prompted us to postulate that whiteflies also vector BaCV.

3.3. *B. tabaci MEAM1 Transmit BaCV-Luz to Common Beans, Cowpea, and Soybean*

The known vectors of cytorhabdoviruses are insects belonging to the families Aphididae (aphids), Delphacidae (planthoppers), and Cicadellidae (leafhoppers). In general, when the vectors are known, there is a highly specific relationship, and only one (or a few related) types or species of a vector are capable of transmitting a given virus. Thus, it is possible to establish a strong correlation between viral detection and the presence of its vector in the field [3,49].

The field in Luziânia, GO, where the bean plants were initially sampled, was densely infested by whiteflies. The genotyping of these whiteflies by PCR-RFLP confirmed their identity as *B. tabaci* MEAM1, the prevalent whitefly species in Central Brazil [50]. To evaluate whether whiteflies also transmit BaCV, we used a BaCV (plus BGMV and CPMMV) field-collected infected bean plant as inoculum source and the flies carried by this plant in a free choice transmission assay (Figure 3). After a 14 day exposition time to the whitefly feeding, 100% (*n* = 48) of the bean plants 'Pérola', 'Jalo', and 'BRS FC 401 RMD' tested positive for the presence of BaCV RNA by RT-PCR. Whitefly adults collected at the same time as bean leaves were also positive for BaCV and CPMMV (Figure S2). All the plants also contained CPMMV, and the susceptible cultivars were positive for BGMV, suggesting that all three viruses were simultaneously transmitted by whiteflies. The mild mottling symptoms detected in 'BRS FC 401 RMD' plants ~50 days after introduction into the cage (Figure 7) resemble those described for CPMMV infection [51]. These results indicate that BaCV could be transmitted at high rates to three different bean cultivars by whiteflies *B. tabaci* MEAM1. The BaCV-Luz isolate was maintained by whitefly-mediated periodical transmission to healthy bean plants for 18 months. Altogether, during this period, 83 plants were exposed to potentially viruliferous flies in the cage, and 72 became infected by BaCV, an overall transmission rate of ~87%.

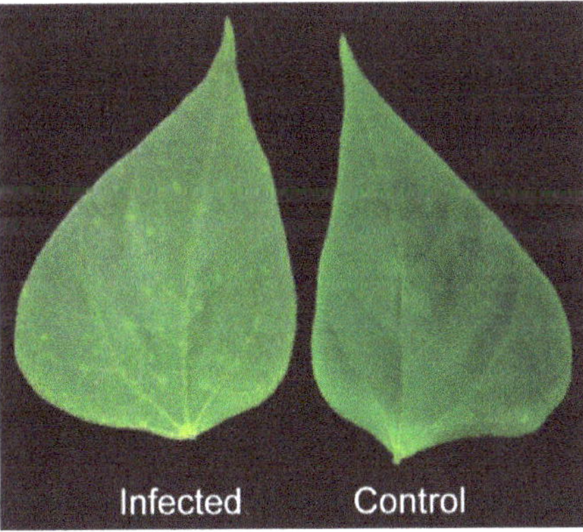

Figure 7. Common bean 'BRS FC 401 RMD' leaves from whitefly mediated BaCV and CPMMV transmission. Mild mottling and chlorotic spots in a leaf of an infected plant at ~50 days after inoculation and the leaf of a non-infected control plant.

A second experiment was conducted to confirm the capacity of whiteflies to transmit BaCV to common bean as well as soybean and cowpea (Figure 3). Using 25 to 30 adult *B. tabaci* MEAM1 per plant and both AAP and IAP of 7 days, BaCV transmission was achieved in 75% of the common beans, 50% of cowpeas, and 25% of the soybean plants (Table 2). All control plants exposed to non-viruliferous insects tested negative for BaCV (and CPMMV). With these results, it was also possible to experimentally extend the host range of BaCV to cowpea and soybean (Table 2). The second experiment showed a transmission efficiency rate lower than the pilot test, in which transmission occurred to 100% of the plants. The reduction in the efficiency of BaCV transmission may be related to the IAP, which was shorter than in the first experiment. Besides, the number of adult insects feeding on each plant was limited to a maximum of 30, while in the first test, the whiteflies were free to feed so that each plant may have received a larger number of viral particles.

Table 2. BaCV transmission by *B. tabaci* MEAM1.

Plant	Cultivar	Positive/Total	Transmission Rate (%)
P. vulgaris (common bean)	'BRSFC 401 RMD'	3/4	75
V. unguiculata (cowpea)	'BRS Imponente'	2/4	50
G. max (soybean)	'BR16'	1/4	25

Furthermore, the whiteflies' ages were not synchronized in the initial trials, and the insect stage of the life cycle may influence the efficiency of BaCV transmission. Soybean is an economically important cash crop in Brazil and is also susceptible to CPMMV [52] and BGMV [53]. Cultivation cycle of soybean and common bean overlap in many areas of Brazil. In these areas, soybeans may act as an inoculum source of these three whitefly-borne viruses to bean crop hasting multiple virus epidemics. Our future investigation should focus on the study of field infection of soybean plants by BaCV. *B. tabaci* is a complex containing approximately 40 cryptic species with similar morphology but differing in the genetics, behavior, efficiency as a virus vector, and in the colonization by endosymbionts [50,54,55]. *B. tabaci* has a large number of hosts, more than 500 plant species, cultivated or not, in tropical and subtropical regions. *B. tabaci* is considered a super vector since it transmits over 300 plant viruses including begomoviruses (family *Geminiviridae*), criniviruses (family *Closteroviridae*), torradoviruses (family *Secoviridae*), ipomoviruses (family *Potyviridae*), and the carlaviruses CPMMV and melon yellowing-associated virus (MYaV) (family *Betaflexiviridae* [56,57]). Recently, two poleroviruses (family *Luteoviridae*) were also shown to be transmitted by *B. tabaci* [58,59].

Our transmission results demonstrate that whiteflies, in this case the species *B. tabaci* MEAM1, are vectors of the cytorhabdovirus BaCV in Brazil, highlighting the importance of whiteflies as plant virus vectors and emphasizing their designation as super vectors [56]. Moreover, our results expand the whiteflies' attributes as vectors, including the cytorhabdovirus group, to the list of viruses they can transmit.

Despite whiteflies being successful plant virus vectors, the transmission efficiency may vary depending on the virus, virus isolate, host plant, whitefly species and biology, and virus and whitefly population's geographical origin [57]. In our study, *B. tabaci* MEAM1 was very efficient in transmitting BaCV, especially to bean plants. *B. tabaci* MEAM1 predominates as a vector of various plant viruses in Brazil [50]. However, other whitefly species such as *B. tabaci* Mediterranean (MED) and *B. tabaci* New World (NW) are also present in more restricted areas of Southeastern and Southern States. Whether whiteflies MED and NW can transmit BaCV to beans or other crops remains to be investigated.

It is also necessary to investigate if *B. tabaci* MEAM1 is the vector of PpVE to papaya plants in Ecuador and if this virus infects papayas in Brazil. Cornejo-Franco, et al. [60] mention that *B. tabaci* is a major pest of papayas in Ecuador and is the vector of papaya virus Q (an umbra-like virus). Thus, this whitefly species may also be the vector PpVE to papayas. We have tested a limited number of papaya plants ($n = 27$) collected in the Distrito Federal and the State of Espirito Santo for the presence of BaCV with negative results. Nine whitefly species, including *B. tabaci* MEAM1, were already identified on papaya trees in Brazil, but *Trialeurodes variabilis* is the primary species associated with this fruit crop, and *B. tabaci* MEAM1 has a limited occurrence in papayas in the country [61]. Considering that BaCV and PpVE are frequently detected in mixed infections, it is important to investigate if this group of viruses is transmitted alone by *B. tabaci* MEAM1 or if a helper virus is required.

3.4. BaCV-Luz Detection in Plants and Whiteflies

Depending on the taxonomic group they belong to, viruses are transmitted by their whitefly vectors by different transmission modes [57]. The carlavirus CPMMV is reported to be stylet borne and transmitted in a nonpersistent mode [52]. While criniviruses, torradoviruses, and ipomoviruses are semipersistent viruses and foregut borne, the mode of transmission of poleroviruses by *B. tabaci* is still unknown [56,57]. Begomoviruses are transmitted by different species in the *B. tabaci* complex in a persistent circulative manner. However, at least for one begomovirus, tomato yellow leaf curl virus (TYLCV), there is evidence that it replicates in the whitefly [62–65] and that replication takes place mainly in the salivary glands [66].

Plant rhabdoviruses are transmitted by their arthropod vectors in a persistent, circulative-propagative manner. After the acquisition, viruliferous insects transmit plant rhabdoviruses for their entire lives. In plants, rhabdoviruses infect, replicate, and accumulate in a variety of tissues, including leaf epidermis and mesophyll, phloem tissues, and roots [5,67]. In their insect vectors, plant rhabdoviruses infect gut cells, muscle cells, nervous tissue, hemocytes, tracheae, and salivary glands [3,5,28].

BaCV RNA corresponding to N, P, P3, P4, M, G genes were amplified from BaCV-infected bean leaves and potentially viruliferous whiteflies collected in infected plants. Fragments with sizes according to their predicted ORFs were amplified (Figure 8a,b), except for L, probably because of its large size. Amplicons corresponding to *act11* and *RbcS* were only amplified from bean cDNA, and *v-ATPase* and *RpL9* from whiteflies (Figure 8a,b).

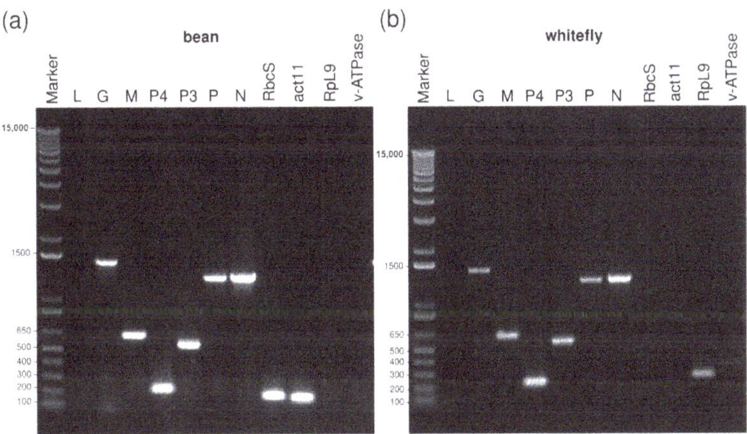

Figure 8. BaCV detection in common bean leaf and whiteflies (*B. tabaci* MEAM1). BaCV RNA corresponding to G, M, P4, P3, P, N genes were detected by PCR using specific primers. Internal controls include the plant *RbcS* and *act11* and the whitefly *RpL9* and *v-ATPase*. (**a**) RT-PCR products obtained from RNA extracted from common bean. (**b**) RT-PCR products obtained from RNA extracted from viruliferous whiteflies.

We have used transmission electron microscopy (TEM) to visualize BaCV accumulation in bean and whitefly tissues. Electron microscopic examination of BaCV-infected bean leaves revealed bacilliform particles typical of rhabdoviruses in parenchymal cells. Longitudinal and cross-sectioned particles were located in the cell cytoplasm, commonly at the periphery of an electron-lucent mass of coiled filamentous material, believed to be the viroplasm, where the virus replicates (Figure 9a–d) [5,68]. The BaCV bacilliform particles seemed rather scarce in the observed bean tissues. Presumed cytorhabdovirus virions were found in only one of the four examined leaf samples, though all plant samples tested positive for BaCV by RT-PCR.

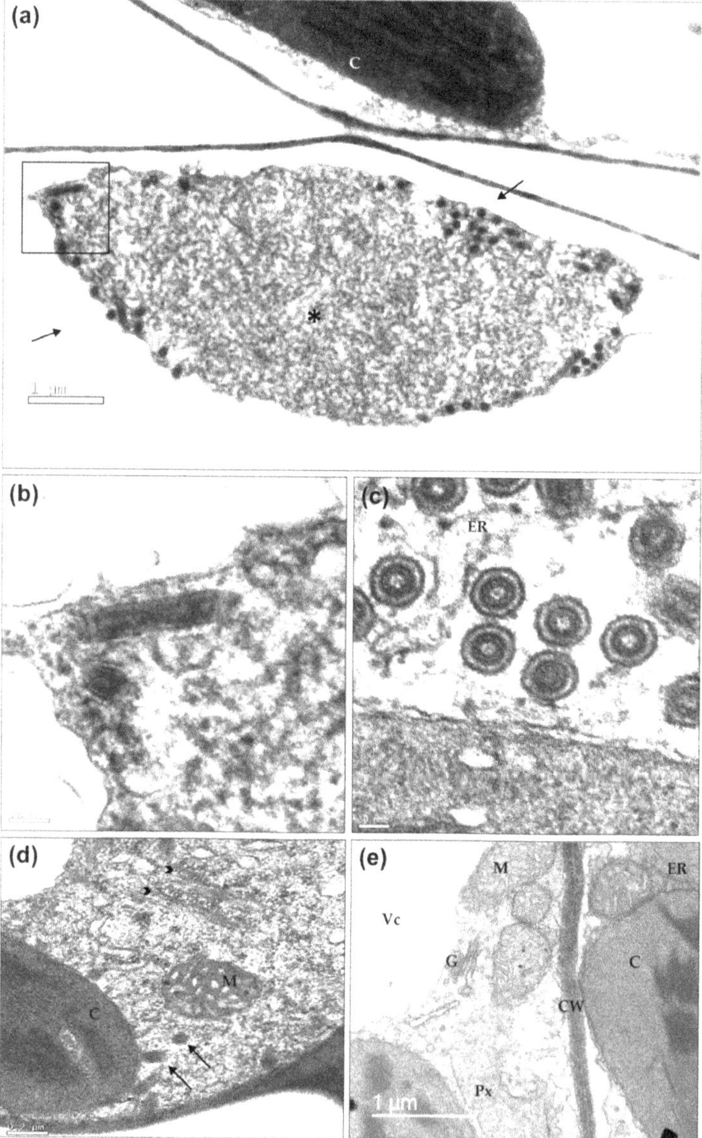

Figure 9. Transmission electron micrographs of bean leaf infected by BaCV. (**a**) Overview of a viroplasm formed by coiled filamentous material (*) in the cytoplasm of a parenchymal cell. Typical rhabdovirus particles are present in the periphery of the viroplasm (arrows). (**b**) Details of the marked square with longitudinally-sectioned particles are depicted. (**c**) Cross-sections of BaCV particles show the internal and cylindrical nucleoprotein core, and the outer viral membrane, and also that virions are within a cavity of the endoplasmic reticulum. (**d**) Spongy parenchymal cell dually infected by BaCV and CPMMV. Brush-like aggregates of CPMMV particles (arrowheads) and BaCV in longitudinal and cross-sections (arrows) are visible. (**e**) Palisade parenchyma cells from a healthy bean plant. Chloroplast (C), endoplasmic reticulum (ER), Golgi complex (G), mitochondrion (M), peroxisome (Px), vacuole (Vc), and cell wall (CW).

By contrast, feather-like aggregates of CPMMV particles were readily recognized in all examined bean tissue samples. These in situ observations of BaCV and CPMMV virions presence in dually infected bean leaf corroborate the previous HTS sequencing study in which 8.2% of the total sequence reads from multiple virus-infected bean leaves corresponded to CPMMV whereas only 0.01% mapped to BaCV [6]. In some cases, BaCV and CPMMV virions could be identified infecting the same cell (Figure 9d). Unfortunately, sections of three pairs of head/thorax or abdominal parts of whiteflies that fed in BaCV-infected beans did not yield evidence of BaCV particles or viroplasm in the observed tissues. The insects were tested before fixation, and only head/thorax corresponding to insects that the abdomen tested positive for BaCV or vice-versa was examined. Therefore, albeit BaCV RNA could be detected in the insects by RT-PCR, particles were not localized in any of the tissues examined.

The spatial and temporal distribution of BaCV within the whitefly body could have hindered the localization of particles by TEM. Moreover, plant rhabdoviruses appear to replicate and accumulate at lower levels in insect cells when compared to plant cells [3,6,69]. BaCV seems to accumulate at low levels, even in bean plants. Therefore, assessment of dissected organs such as midguts or salivary glands could facilitate the visualization of BaCV in whiteflies either by TEM or by confocal microscopy using immunofluorescence or in situ hybridization. The occurrence of BaCV vRNA, cRNA, and mRNA in whiteflies will be investigated in future studies.

4. Conclusions

Our results show that BaCV is efficiently transmitted, in experimental conditions, by *B. tabaci* MEAM1 to three bean cultivars grown in Brazil, and with lower efficiency to cowpea and soybean. It remains to be determined whether BaCV replicates in whiteflies, as observed for other plant-infecting rhabdoviruses in their arthropod vectors. BaCV could not be singly isolated for pathogenicity tests, identification of the induced symptoms, and the transmission assay. Moreover, BaCV was detected in five out of seven Brazilian states evaluated. Besides BaCV in Brazil (this study, [6]) and PpVE in Ecuador [7], similar virus sequences were recorded from whiteflies samples in India [70], beans from Tanzania [71], and citrus, passion fruit, and paper bush in China [72], implying that other isolates/strains of BaCV/PpVE or related rhabdoviruses that are also transmitted by whiteflies might exist in other continents.

Supplementary Materials: The following are available online at http://www.mdpi.com/1999-4915/12/9/1028/s1, Figure S1: RT-PCR and PCR detection of bean-infecting viruses, Figure S2: RT-PCR detection of BaCV and CPMMV, Table S1: Primers used in this study, Table S2: Accession number of cytorhabdoviruses sequences in GenBank and their known vectors, Table S3: Nucleotide and amino acid sequence identities (%) of BaCV-Luz ORFs compared with PpEV strains.

Author Contributions: Conceptualization, B.P.-L., R.C.P.-C., S.G.R. and F.L.M.; Data curation, B.P.-L. and F.L.M.; Formal analysis, B.P.-L., R.S.F. and F.L.M.; Funding acquisition, A.V., S.G.R. and F.L.M.; Investigation, B.P.-L., D.M.T.A.-F., E.W.K., A.H.V., M.T.G. and R.S.F.; Methodology, B.P.-L., E.W.K. and F.L.M.; Project administration, J.C.F., S.G.R. and F.L.M.; Resources, J.C.F., E.F.M.A., S.G.R. and F.L.M.; Supervision, R.C.P.-C., S.G.R. and F.L.M.; Visualization, B.P.-L., C.L. and F.L.M.; Writing—original draft, B.P.-L., S.G.R. and F.L.M.; Writing—review & editing, B.P.-L., R.C.P.-C., D.M.T.A.-F., E.W.K., A.H.V., C.L., R.S.F., J.C.F., A.V., S.G.R. and F.L.M. All authors have read and agreed to the published version of the manuscript.

Funding: This research was funded by grants from Embrapa, Conselho Nacional de Desenvolvimento Científico e Tecnológico—CNPq and Fundação de Apoio à Pesquisa do Distrito Federal—FAP/DF. B.P.-L. and A.H.V. are supported by scholarships from Coordenação de Aperfeiçoamento de Pessoal de Nível Superior—CAPES. D.M.T.A.-F. and M.T.G. received fellowships from CNPq.

Acknowledgments: We thank Rafael N. Lemos, Rodrigo Carniel, Luis C. Faria, Mônica Neli, and Marlonni Marastoni for helping in sample collection or sharing plant material; Mário Saraiva for technical support and assistance in the lab administration.

Conflicts of Interest: The authors declare no conflict of interest. The funders, including Embrapa, had no role in the design of the study; in the collection, analyses, or interpretation of data; in the writing of the manuscript, or in the decision to publish the results.

References

1. Freitas-Astua, J.; Dietzgen, R.G.; Walker, P.J.; Blasdell, K.R.; Breyta, R.; Fooks, A.R.; Kondo, H.; Kurath, G.; Kuzmin, I.V.; Stone, D.M.; et al. Split the genus Nucleorhabdovirus, creating three new genera (Alphanucleorhabdovirus, Betanucleorhabdovirus and Gammanucleorhabdovirus) comprising sixteen species, including six new species, in the family Rhabdoviridae. *Approved ICTV Proposal.* 2019. Available online: https://ictv.global/ICTV/proposals/2019.031M.zip (accessed on 18 May 2020).
2. Walker, P.J.; Blasdell, K.R.; Calisher, C.H.; Dietzgen, R.G.; Kondo, H.; Kurath, G.; Longdon, B.; Stone, D.M.; Tesh, R.B.; Tordo, N.; et al. ICTV Virus Taxonomy Profile: Rhabdoviridae. *J. Gen. Virol.* **2018**, *99*, 447–448. [CrossRef]
3. Ammar, E.D.; Tsai, C.W.; Whitfield, A.E.; Redinbaugh, M.G.; Hogenhout, S.A. Cellular and molecular aspects of rhabdovirus interactions with insect and plant hosts. *Annu. Rev. Entomol.* **2009**, *54*, 447–468. [CrossRef] [PubMed]
4. Dietzgen, R.G.; Kondo, H.; Goodin, M.M.; Kurath, G.; Vasilakis, N. The family Rhabdoviridae: Mono- and bipartite negative-sense RNA viruses with diverse genome organization and common evolutionary origins. *Virus Res.* **2017**, *227*, 158–170. [CrossRef]
5. Whitfield, A.E.; Huot, O.B.; Martin, K.M.; Kondo, H.; Dietzgen, R.G. Plant rhabdoviruses-their origins and vector interactions. *Curr. Opin. Virol.* **2018**, *33*, 198–207. [CrossRef]
6. Alves-Freitas, D.M.T.; Pinheiro-Lima, B.; Faria, J.C.; Lacorte, C.; Ribeiro, S.G.; Melo, F.L. Double-Stranded RNA High-Throughput Sequencing Reveals a New Cytorhabdovirus in a Bean Golden Mosaic Virus-Resistant Common Bean Transgenic Line. *Viruses* **2019**, *11*, 90. [CrossRef]
7. Medina-Salguero, A.X.; Cornejo-Franco, J.F.; Grinstead, S.; Mollov, D.; Mowery, J.D.; Flores, F.; Quito-Avila, D.F. Sequencing, genome analysis and prevalence of a cytorhabdovirus discovered in Carica papaya. *PLoS ONE* **2019**, *14*, e0215798. [CrossRef]
8. Freitas-Astua, J.; Dietzgen, R.G.; Walker, P.J.; Blasdell, K.R.; Breyta, R.; Fooks, A.R.; Kondo, H.; Kurath, G.; Kuzmin, I.V.; Stone, D.M.; et al. Create twelve new species in the genus Cytorhabdovirus, family Rhabdoviridae. *Approved ICTV Proposal.* 2019. Available online: https://ictv.global/ICTV/proposals/2019.030M.zip (accessed on 18 May 2020).
9. International Committee on Taxonomy of Viruses—ICTV Online. Available online: https://talk.ictvonline.org/ictv-reports/ictv_online_report/negative-sense-rna-viruses/mononegavirales/w/rhabdoviridae (accessed on 18 May 2020).
10. Instituto Brasileiro de Geografia e Estatística—IBGE. Available online: https://www.ibge.gov.br (accessed on 24 July 2020).
11. Doyle, J.J.; Doyle, J.L. A rapid DNA isolation procedure for small quantities of fresh leaf tissue. *Phytochem. Bull.* **1987**, *19*, 11–15.
12. Nicolini, C.; Pio-Ribeiro, G.; Andrade, G.P.; Melo, F.L.; Oliveira, V.C.; Guimaraes, F.C.; Resende, R.O.; Kitajima, E.W.; Rezende, J.A.; Nagata, T. A distinct tymovirus infecting Cassia hoffmannseggii in Brazil. *Virus Genes* **2012**, *45*, 190–194. [CrossRef]
13. Schuster, D.; Buchman, G.; Rashtchian, A. A simple and efficient method for amplification of cDNA ends using 5′ RACE. *Focus* **1992**, *14*, 46–52.
14. Pinheiro-Lima, B.; Alves-Freitas, D.M.T.; Melo, F.L.; Carvalho, R.; de Faria, J.C.; Ribeiro, S.G. Characterization of a new whitefly-transmitted (Bemisia tabaci MEAM 1) cytorhabdovirus infecting common bean. *Congr. Nac. Pesqui. Feijão* **2017**, *12*, 52.
15. Lamas, N.S.; Matos, V.O.R.L.; Alves-Freitas, D.M.T.; Melo, F.L.; Costa, A.F.; Faria, J.C.; Ribeiro, S.G. Occurrence of Cowpea mild mottle virus in Common Bean and Associated Weeds in Northeastern Brazil. *Plant Dis.* **2017**, *101*, 1828. [CrossRef]
16. Bonfim, K.; Faria, J.C.; Nogueira, E.O.; Mendes, E.A.; Aragao, F.J. RNAi-mediated resistance to Bean golden mosaic virus in genetically engineered common bean (Phaseolus vulgaris). *Mol. Plant Microbe Interact* **2007**, *20*, 717–726. [CrossRef] [PubMed]
17. Bosco, D.; Loria, A.; Sartor, C.; Cenis, J. PCR-RFLP Identification of Bemisia tabaci biotypes in the Mediterranean Basin. *Phytoparasitica* **2006**, *34*, 243–251. [CrossRef]

18. Kearse, M.; Moir, R.; Wilson, A.; Stones-Havas, S.; Cheung, M.; Sturrock, S.; Buxton, S.; Cooper, A.; Markowitz, S.; Duran, C.; et al. Geneious Basic: An integrated and extendable desktop software platform for the organization and analysis of sequence data. *Bioinformatics* **2012**, *28*, 1647–1649. [CrossRef]

19. Altschul, S.F.; Gish, W.; Miller, W.; Myers, E.W.; Lipman, D.J. Basic local alignment search tool. *J. Mol. Biol.* **1990**, *215*, 403–410. [CrossRef]

20. Katoh, K.; Rozewicki, J.; Yamada, K.D. MAFFT online service: Multiple sequence alignment, interactive sequence choice and visualization. *Brief. Bioinform.* **2019**, *20*, 1160–1166. [CrossRef]

21. Trifinopoulos, J.; Nguyen, L.T.; von Haeseler, A.; Minh, B.Q. W-IQ-TREE: A fast online phylogenetic tool for maximum likelihood analysis. *Nucleic Acids Res.* **2016**, *44*, W232–W235. [CrossRef]

22. Shimodaira, H. An approximately unbiased test of phylogenetic tree selection. *Syst. Biol.* **2002**, *51*, 492–508. [CrossRef]

23. Gerlt, J.A.; Bouvier, J.T.; Davidson, D.B.; Imker, H.J.; Sadkhin, B.; Slater, D.R.; Whalen, K.L. Enzyme Function Initiative-Enzyme Similarity Tool (EFI-EST): A web tool for generating protein sequence similarity networks. *Biochim. Biophys. Acta* **2015**, *1854*, 1019–1037. [CrossRef]

24. Shannon, P.; Markiel, A.; Ozier, O.; Baliga, N.S.; Wang, J.T.; Ramage, D.; Amin, N.; Schwikowski, B.; Ideker, T. Cytoscape: A software environment for integrated models of biomolecular interaction networks. *Genome Res.* **2003**, *13*, 2498–2504. [CrossRef]

25. Muhire, B.M.; Varsani, A.; Martin, D.P. SDT: A virus classification tool based on pairwise sequence alignment and identity calculation. *PLoS ONE* **2014**, *9*, e108277. [CrossRef] [PubMed]

26. Faria, J.C.; Aragão, F.J.L.; Souza, T.L.P.O.; Quintela, E.D.; Kitajima, E.W.; Ribeiro, S.G. Golden mosaic of common beans in Brazil: Management with a transgenic approach. *APS Features* **2016**, *10*, 1–14.

27. Orfanidou, C.G.; Pappi, P.G.; Efthimiou, K.E.; Katis, N.I.; Maliogka, V.I. Transmission of Tomato chlorosis virus (ToCV) by Bemisia tabaci Biotype Q and Evaluation of Four Weed Species as Viral Sources. *Plant Dis.* **2016**, *100*, 2043–2049. [CrossRef] [PubMed]

28. Cao, Q.; Xu, W.Y.; Gao, Q.; Jiang, Z.H.; Liu, S.Y.; Fang, X.D.; Gao, D.M.; Wang, Y.; Wang, X.B. Transmission Characteristics of Barley Yellow Striate Mosaic Virus in Its Planthopper Vector Laodelphax striatellus. *Front. Microbiol.* **2018**, *9*, 1419. [CrossRef]

29. Borges, A.; Tsai, S.M.; Caldas, D.G. Validation of reference genes for RT-qPCR normalization in common bean during biotic and abiotic stresses. *Plant Cell Rep.* **2012**, *31*, 827–838. [CrossRef]

30. Knight, M.R.; Jenkins, G.I. Genes encoding the small subunit of ribulose 1,5-bisphosphate carboxylase/oxygenase in *Phaseolus vulgaris* L.: Nucleotide sequence of cDNA clones and initial studies of expression. *Plant Mol. Biol.* **1992**, *18*, 567–579. [CrossRef] [PubMed]

31. Upadhyay, S.K.; Chandrashekar, K.; Thakur, N.; Verma, P.C.; Borgio, J.F.; Singh, P.K.; Tuli, R. RNA interference for the control of whiteflies (Bemisia tabaci) by oral route. *J. Biosci.* **2011**, *36*, 153–161. [CrossRef] [PubMed]

32. Alves-Freitas, D.M.T.; Melo, F.L.; Faria, J.C.; Ribeiro, S.G. dsRNA deep sequencing reveals five viral species in common beans. *J. Braz. Soc. Virol.* **2016**, *20*, 43–44.

33. Moreno, A.B.; Lopez-Moya, J.J. When Viruses Play Team Sports: Mixed Infections in Plants. *Phytopathology* **2020**, *110*, 29–48. [CrossRef] [PubMed]

34. Gautam, S.; Gadhave, K.R.; Buck, J.W.; Dutta, B.; Coolong, T.; Adkins, S.; Srinivasan, R. Virus-virus interactions in a plant host and in a hemipteran vector: Implications for vector fitness and virus epidemics. *Virus Res.* **2020**, *286*, 198069. [CrossRef] [PubMed]

35. Francki, R.I.B.; Randles, J.W.; Dietzgen, R.G. Lettuce necrotic yellows virus. *AAB Descr. Plant Viruses* **1989**, *343*, 5.

36. Jones, S.; McGavin, W.; MacFarlane, S. The complete sequences of two divergent variants of the rhabdovirus raspberry vein chlorosis virus and the design of improved primers for virus detection. *Virus Res.* **2019**, *265*, 162–165. [CrossRef]

37. Sylvester, E.S.; Richardson, J.; Frazier, N.W. Serial passage of strawberry crinkle virus in the aphid Chaetosiphon jacobi. *Virology* **1974**, *59*, 301–306. [CrossRef]

38. Franova, J.; Pribylova, J.; Koloniuk, I. Molecular and Biological Characterization of a New Strawberry Cytorhabdovirus. *Viruses* **2019**, *11*, 982. [CrossRef]

39. Li, C.X.; Shi, M.; Tian, J.H.; Lin, X.D.; Kang, Y.J.; Chen, L.J.; Qin, X.C.; Xu, J.; Holmes, E.C.; Zhang, Y.Z. Unprecedented genomic diversity of RNA viruses in arthropods reveals the ancestry of negative-sense RNA viruses. *eLife* **2015**, *4*. [CrossRef] [PubMed]

40. Conti, M.; Appiano, A. Barley yellow striate mosaic virus and associated viroplasms in barley cells. *J. Gen. Virol.* **1973**, *21*, 315–322. [CrossRef]

41. Gollifer, D.E.; Jackson, G.V.H.; Dabek, A.J.; Plumb, R.T.; May, Y.Y. The occurrence and transmission of viruses of edible aroids in the Solomon Islands and the Southwest Pacific. *Pestic. Artic. News Summ.* **1977**, *23*, 171–177. [CrossRef]

42. Maurino, M.F.; Laguna, G.; Giolitti, F.; Nome, C.; Pecci, M.P.G. First Ocurrence of a Rhabdovirus Infecting Maize in Argentina. *Plant Dis.* **2012**, *96*, 1383. [CrossRef] [PubMed]

43. Maurino, F.; Dumon, A.D.; Llauger, G.; Alemandri, V.; de Haro, L.A.; Mattio, M.F.; Del Vas, M.; Laguna, I.G.; Gimenez Pecci, M.P. Complete genome sequence of maize yellow striate virus, a new cytorhabdovirus infecting maize and wheat crops in Argentina. *Arch. Virol.* **2018**, *163*, 291–295. [CrossRef] [PubMed]

44. Toriyama, S. Purification and some properties of northern cereal mosaic virus. *Uirusu* **1972**, *22*, 114–124. [CrossRef]

45. Toriyama, S. Northern cereal mosaic. In *Viruses and Virus Diseases of Poaceae (Gramineae)*; Lapierre, H., Signoret, P.-A., Eds.; INRA: Paris, France, 2004; pp. 578–580.

46. Yang, X.; Huang, J.; Liu, C.; Chen, B.; Zhang, T.; Zhou, G. Rice Stripe Mosaic Virus, a Novel Cytorhabdovirus Infecting Rice via Leafhopper Transmission. *Front. Microbiol.* **2016**, *7*, 2140. [CrossRef] [PubMed]

47. Bejerman, N.; Dietzgen, R.G. Letter to the Editor: Bean-Associated Cytorhabdovirus and Papaya Cytorhabdovirus are Strains of the Same Virus. *Viruses* **2019**, *11*, 230. [CrossRef] [PubMed]

48. Dietzgen, R.G.; Bejerman, N.E.; Goodin, M.M.; Higgins, C.M.; Huot, O.B.; Kondo, H.; Martin, K.M.; Whitfield, A.E. Diversity and epidemiology of plant rhabdoviruses. *Virus Res.* **2020**, *281*, 197942. [CrossRef] [PubMed]

49. Mann, K.S.; Dietzgen, R.G. Plant rhabdoviruses: New insights and research needs in the interplay of negative-strand RNA viruses with plant and insect hosts. *Arch. Virol.* **2014**, *159*, 1889–1900. [CrossRef]

50. de Moraes, L.A.; Muller, C.; Bueno, R.; Santos, A.; Bello, V.H.; De Marchi, B.R.; Watanabe, L.F.M.; Marubayashi, J.M.; Santos, B.R.; Yuki, V.A.; et al. Distribution and phylogenetics of whiteflies and their endosymbiont relationships after the Mediterranean species invasion in Brazil. *Sci. Rep.* **2018**, *8*, 14589. [CrossRef]

51. Carvalho, S.L.; Nagata, T.; Junqueira, B.R.; Zanardo, L.G.; Paiva, A.C.; Carvalho, C.M. Construction of a full length infectious cDNA clone of Cowpea mild mottle virus. *Virus Genes* **2017**, *53*, 137–140. [CrossRef]

52. Zanardo, L.G.; Carvalho, C.M. Cowpea mild mottle virus (Carlavirus, Betaflexiviridae): A review. *Trop. Plant Pathol.* **2017**, *42*, 417–430. [CrossRef]

53. Fernandes, F.R.; Cruz, A.R.; Faria, J.C.; Zerbini, F.M.; Aragao, F.J. Three distinct begomoviruses associated with soybean in central Brazil. *Arch. Virol.* **2009**, *154*, 1567–1570. [CrossRef]

54. De Barro, P.J.; Liu, S.S.; Boykin, L.M.; Dinsdale, A.B. Bemisia tabaci: A statement of species status. *Annu. Rev. Entomol.* **2011**, *56*, 1–19. [CrossRef]

55. Lee, W.; Park, J.; Lee, G.S.; Lee, S.; Akimoto, S. Taxonomic status of the Bemisia tabaci complex (Hemiptera: Aleyrodidae) and reassessment of the number of its constituent species. *PLoS ONE* **2013**, *8*, e63817. [CrossRef]

56. Gilbertson, R.L.; Batuman, O.; Webster, C.G.; Adkins, S. Role of the Insect Supervectors Bemisia tabaci and Frankliniella occidentalis in the Emergence and Global Spread of Plant Viruses. *Annu. Rev. Virol.* **2015**, *2*, 67–93. [CrossRef] [PubMed]

57. Fiallo-Olive, E.; Pan, L.L.; Liu, S.S.; Navas-Castillo, J. Transmission of Begomoviruses and Other Whitefly-Borne Viruses: Dependence on the Vector Species. *Phytopathology* **2020**, *110*, 10–17. [CrossRef] [PubMed]

58. Ghosh, S.; Kanakala, S.; Lebedev, G.; Kontsedalov, S.; Silverman, D.; Alon, T.; Mor, N.; Sela, N.; Luria, N.; Dombrovsky, A.; et al. Transmission of a New Polerovirus Infecting Pepper by the Whitefly Bemisia tabaci. *J. Virol.* **2019**, *93*. [CrossRef] [PubMed]

59. Costa, T.M.; Inoue-Nagata, A.K.; Vidal, A.H.; Ribeiro, S.G.; Nagata, T. The recombinant isolate of cucurbit aphid-borne yellows virus from Brazil is a polerovirus transmitted by whiteflies. *Plant Pathol.* **2020**, *69*, 1042–1050. [CrossRef]

60. Cornejo-Franco, J.F.; Alvarez-Quinto, R.A.; Quito-Avila, D.F. Transmission of the umbra-like Papaya virus Q in Ecuador and its association with meleira-related viruses from Brazil. *Crop Prot.* **2018**, *110*, 99–102. [CrossRef]

61. Martins, D.S.; Lima, A.F.; Fornazier, M.J.; Barcellos, B.D.; Queiroz, R.B.; Fanton, C.J.; Zanuncio-Junior, J.S.; Fornazier, D.L. Whiteflies (Hemiptera: Aleyrodidae) Associated with papaya (Carica papaya L.). *Rev. Científica Intelletto* **2016**, *2*, 78–86.

62. Sinisterra, X.H.; McKenzie, C.L.; Hunter, W.B.; Powell, C.A.; Shatters, R.G. Differential transcriptional activity of plant-pathogenic begomoviruses in their whitefly vector (Bemisia tabaci, Gennadius: Hemiptera Aleyrodidae). *J. Gen. Virol.* **2005**, *86*, 1525–1532. [CrossRef]

63. Pakkianathan, B.C.; Kontsedalov, S.; Lebedev, G.; Mahadav, A.; Zeidan, M.; Czosnek, H.; Ghanim, M. Replication of Tomato Yellow Leaf Curl Virus in Its Whitefly Vector, Bemisia tabaci. *J. Virol.* **2015**, *89*, 9791–9803. [CrossRef]

64. Wang, L.L.; Wang, X.R.; Wei, X.M.; Huang, H.; Wu, J.X.; Chen, X.X.; Liu, S.S.; Wang, X.W. The autophagy pathway participates in resistance to tomato yellow leaf curl virus infection in whiteflies. *Autophagy* **2016**, *12*, 1560–1574. [CrossRef]

65. Czosnek, H.; Ghanim, M.; Morin, S.; Rubinstein, G.; Fridman, V.; Zeidan, M. Whiteflies: Vectors, and victims (?), of geminiviruses. *Adv. Virus Res.* **2001**, *57*, 291–322. [CrossRef]

66. He, Y.Z.; Wang, Y.M.; Yin, T.Y.; Fiallo-Olive, E.; Liu, Y.Q.; Hanley-Bowdoin, L.; Wang, X.W. A plant DNA virus replicates in the salivary glands of its insect vector via recruitment of host DNA synthesis machinery. *Proc. Natl. Acad. Sci. USA* **2020**, *117*, 16928–16937. [CrossRef]

67. Jackson, A.O.; Dietzgen, R.G.; Goodin, M.M.; Bragg, J.N.; Deng, M. Biology of plant rhabdoviruses. *Annu. Rev. Phytopathol.* **2005**, *43*, 623–660. [CrossRef] [PubMed]

68. Jackson, A.O.; Dietzgen, R.G.; Goodin, M.M.; Li, Z. Development of Model Systems for Plant Rhabdovirus Research. *Adv. Virus Res.* **2018**, *102*, 23–57. [CrossRef] [PubMed]

69. Ammar, E.D.; Hogenhout, S.A. A neurotropic route for Maize mosaic virus (Rhabdoviridae) in its planthopper vector Peregrinus maidis. *Virus Res.* **2008**, *131*, 77–85. [CrossRef]

70. Upadhyay, S.K.; Sharma, S.; Singh, H.; Dixit, S.; Kumar, J.; Verma, P.C.; Chandrashekar, K. Whitefly genome expression reveals host-symbiont interaction in amino acid biosynthesis. *PLoS ONE* **2015**, *10*, e0126751. [CrossRef]

71. Mwaipopo, B.; Nchimbi-Msolla, S.; Njau, P.J.R.; Mark, D.; Mbanzibwa, D.R. Comprehensive Surveys of Bean common mosaic virus and Bean common mosaic necrosis virus and Molecular Evidence for Occurrence of Other Phaseolus vulgaris Viruses in Tanzania. *Plant Dis.* **2018**, *102*, 2361–2370. [CrossRef] [PubMed]

72. Zhang, S.; Huang, A.; Zhou, X.; Li, Z.; Dietzgen, R.G.; Zhou, C.Y.; Cao, M. Natural defect of a plant rhabdovirus glycoprotein gene: A case study of virus-plant co-evolution. *Phytopathology* **2020**. [CrossRef]

Article

Can Winged Aphid Abundance Be a Predictor of Cucurbit Aphid-Borne Yellows Virus Epidemics in Melon Crop?

Alexandra Schoeny *, Loup Rimbaud, Patrick Gognalons, Grégory Girardot, Pauline Millot, Karine Nozeran, Catherine Wipf-Scheibel and Hervé Lecoq

INRAE, Pathologie Végétale, F-84140 Montfavet, France; loup.rimbaud@inrae.fr (L.R.); patrick.gognalons@inrae.fr (P.G.); gregory.girardot@inrae.fr (G.G.); pauline.millot@inrae.fr (P.M.); karine.nozeran@inrae.fr (K.N.); catherine.rys@inrae.fr (C.W.-S.); rvbalecoq@gmail.com (H.L.)
* Correspondence: alexandra.schoeny@inrae.fr

Received: 31 July 2020; Accepted: 17 August 2020; Published: 20 August 2020

Abstract: Aphid-borne viruses are frequent yield-limiting pathogens in open field vegetable crops. In the absence of curative methods, virus control relies exclusively on measures limiting virus introduction and spread. The efficiency of control measures may greatly benefit from an accurate knowledge of epidemic drivers, in particular those linked with aphid vectors. Field experiments were conducted in southeastern France between 2010 and 2019 to investigate the relationship between the epidemics of cucurbit aphid-borne yellows virus (CABYV) and aphid vector abundance. Winged aphids visiting melon crops were sampled daily to assess the abundance of CABYV vectors (*Aphis gossypii*, *Macrosiphum euphorbiae* and *Myzus persicae*) and CABYV was monitored weekly by DAS-ELISA. Epidemic temporal progress curves were successfully described by logistic models. A systematic search for correlations was undertaken between virus variables including parameters μ (inflection point of the logistic curve) and γ (maximum incidence) and aphid variables computed by aggregating abundances on periods relative either to the planting date, or to the epidemic peak. The abundance of *A. gossypii* during the first two weeks after planting was found to be a good predictor of CABYV dynamics, suggesting that an early control of this aphid species could mitigate the onset and progress of CABYV epidemics in melon crops.

Keywords: *Aphis gossypii*; *Cucumis melo*; cucurbit viruses; disease progress curve; insect trapping; logistic model; Spearman correlation; temporal dynamics

1. Introduction

In France, melon (*Cucumis melo* var. *cantalupensis*) is cultivated in three main production areas: South-East, South-West and Centre-West. South-East produces around 40% of the national production (224,720 t, 11,720 ha in 2019, www.agreste.agriculture.gouv.fr). Open field melon crops are frequently infected by viruses among which is cucurbit aphid-borne yellows virus (CABYV, Polerovirus, Luteoviridae). Since its first report in the 1980s [1], CABYV has been detected in an ever-increasing number of countries [2] and recent surveys indicate that it is becoming prevalent in many cucurbit growing areas [3,4]. It induces typical symptoms of yellowing of the older leaves and impacts yield via flower abortion and reduced number of fruits per plant. It is transmitted in a persistent, circulative, non-propagative manner by a few aphid species (*Aphis gossypii*, *Macrosiphum euphorbiae* and *Myzus persicae*) [5]. In melon crops, *A. gossypii* seems to be largely involved in CABYV epidemics [6,7] but it is still not clear if monitoring *A. gossypii* abundance could be used as a predictor of CABYV epidemics.

In the absence of curative methods, virus control relies exclusively on measures limiting virus introduction and spread (prophylaxis, genetic resistance, vector control, sanitation) [8]. Concerning

CABYV, although resistance genes have been identified in at least two melon accessions [9,10], to our knowledge they have not yet been integrated in commercial cultivars. A recent study showed that the *Vat* gene conferring resistance to the melon aphid *A. gossypii* and the viruses it carries [11] had a significant impact on CABYV epidemics with the mean incidence reduction exceeding 50% in some trials [7]. Still, this effect is far from meeting farmer's expectations. Therefore, complementary measures should also be employed to limit virus introduction and spread in the field. Whatever the control method under consideration (chemical, genetic, biological), its efficiency is likely to be enhanced with an improved knowledge of epidemic drivers, in particular those linked with aphid vectors. Indeed, a good understanding of the relationship between aphid vector abundance and virus epidemic dynamics will certainly help improve (i) our ability to predict future epidemics and (ii) the timeline of control measures intended to prevent the arrival and intra-field movement of aphid vectors as well as the onset and progress of viral diseases.

In this study, we investigated the relationships between aphid population dynamics and CABYV epidemics in melon crops assessed in field experiments conducted in southeastern France between 2010 and 2019. For this, we monitored both aphid populations and viral dynamics, resulting in two distinct datasets. We then looked for potential relationships between numerous variables computed from these two datasets with the overall aim of modelling CABYV epidemics using aphid abundances.

2. Materials and Methods

2.1. Field Experiments

Eleven field experiments were conducted between 2010 and 2019 in Avignon, southeastern France (Table 1): nine at the St Paul experimental station (43°54′53″ N, 4°52′59″ E) and two at the St Maurice experimental station (43°56′49″ N, 4°51′52″ E). The two sites are approximately 4 km apart. Although run in the framework of different projects, all trials involved a Charentais-type melon plot (Charentais T line, susceptible to the melon aphid *A. gossypii*) surrounded by bare soil maintained through mechanical weeding. Seedlings were prepared in an insect-proof greenhouse three weeks before planting. Depending on the trial, plants at the 1–3 leaf stage were planted in late April or late May on dark brown plastic mulch with drip irrigation. Early plantings were protected from wind and cold damage with Agryl P17 fleece (Fiberweb France, Biesheim) for 11–15 days. This row cover also protects from virus contaminations by preventing viruliferous aphids to reach the plants [8]. CABYV is not seed-borne and all plantlets grown in an insect-proof greenhouse or under Agryl P17 may be considered as healthy at the planting or fleece removal stages. The experimental plot comprised 120 to 240 plants (0.5–0.8 m plant spacing) organized in 6 to 16 rows (1.5–2 m row spacing) depending on the trial (Table 1). No insecticides were applied during the trials.

2.2. Plant Sampling and Virus Monitoring

In order to monitor virus dynamics, melon plants were sampled weekly, 8 times during the cropping season. Sample collections generally started one week after planting/fleece removal and investigated 13 to 60% of the crop depending on the trial (Table 1). Sampling plans were designed regardless of the presence of virus symptoms. For the sake of simplification, the nth day after planting or fleece removal will be coded as "Dn" later in the article. Expanding leaves were sampled at each sampling date. It was assumed that information obtained from an expanding leaf accurately reflects the status of the entire plant. CABYV was diagnosed via double antibody sandwich enzyme-linked immunosorbent assay (DAS-ELISA) with a specific polyclonal antiserum [1]. Virus detection was considered positive when the absorbance at 405 nm was greater than 3 times that of the mean of healthy controls.

Table 1. Melon crop and sampling details for field trials conducted in Avignon between 2010 and 2019.

Trial Code	Experimental Site	Planting Date	Number of Plants	Number of Rows	Number of Plants Per Row	Row Spacing (m)	Plant Spacing (m)	Number of Plants Sampled Per Date (Sampling Effort %)
M10	St Paul	28/05/2010	160	8	20	2	0.8	26 (16%)
V11	St Paul	09/05/2011 [a]	120	6	20	2	0.5	24 (20%)
V12	St Paul	11/05/2012 [a]	150	6	25	2	0.5	24 (16%)
V13	St Paul	06/05/2013 [a]	150	6	25	2	0.5	24 (16%)
P11	St Paul	24/05/2011	208	16	13	1.5	0.5	40 (19%)
P12	St Paul	31/05/2012	240	16	15	1.5	0.5	40 (17%)
P13	St Paul	24/05/2013	240	16	15	1.5	0.5	32 (13%)
P14	St Paul	27/05/2014	240	16	15	1.5	0.5	40 (17%)
P15	St Paul	28/05/2015	240	16	15	1.5	0.5	40 (17%)
M18	St Maurice	25/05/2018	160	8	20	1.5	0.5	96 (60%)
M19	St Maurice	28/05/2019	160	8	20	1.5	0.5	96 (60%)

[a] Agryl P17 fleece removal; fleece optimizes plant growth by increasing both air and soil temperatures and reducing wind damage.

2.3. Insect Sampling and Aphid Monitoring

The temporal dynamics of winged aphids visiting the melon crops were established from planting or fleece removal (D0), and lasted until the end of the virus sampling period. Winged insects were sampled at the crop height with a non-biased suction trap [12]. Catches were collected daily, rinsed and stored in 70% ethanol until sorting (aphids vs. other insects) and taxonomic identification of aphids under a stereomicroscope. Aphid abundance datasets used in this study are described in detail in [12]. The analyses focused on the three CABYV vectors reported in the literature: *A. gossypii*, *M. persicae* and *M. euphorbiae*, respectively associated with their Rothamsted Insect Survey (RIS) codes: RIS-181, RIS-322 and RIS-410.

2.4. Computation of Variables Related to Virus Epidemics

For each trial and sampling date, CABYV incidence was calculated as the ratio of the number of infected plants divided by the number of sampled plants. Datasets were standardized in order to facilitate data mining: when missing, incidences at D7, D14, D21, D28, D35, D42, D49 and D56 were estimated by linear interpolation from surrounding sampling dates.

CABYV epidemics were summarized by four "virus" variables. The first one is the area under the disease progress curve (AUDPC) from D0 to D56 calculated according to the formula (1): where y_i represents CABYV incidence, expressed as a percentage, at date D_i.

$$\text{AUDPC}_{56} = \sum_{i=0}^{i=55} \frac{\left[y_i + y_{i+1}\right]}{2} \times (D_{i+1} - D_i) \tag{1}$$

AUDPC_{56} was divided by the total virus monitoring duration (56 days) to calculate the mean incidence over the epidemic. On the basis of their mean incidence, epidemics were categorized as mild (0–20%), intermediate (21–40%), severe (41–60%) or extreme (61–100%) [7].

The three following virus variables are parameters of a logistic equation fitted to incidence data using nonlinear least squares (2):

$$y_{t,k} = \frac{\gamma_k}{1 + e^{-4.\alpha_k.(t-\mu_k)}} \tag{2}$$

where $y_{t,k}$ is the incidence, expressed as a percentage, at time t ($t \in [\![1;56]\!]$) and for trial k ($k \in [\![1;11]\!]$); μ_k is the abscissa of the inflection point for trial k, i.e., the date of the epidemic peak. Low values of μ indicate precocious epidemics while high values are associated with late epidemics; γ_k is the plateau, i.e., the carrying capacity, for trial k. High values of γ indicate global epidemics (high incidence at the end of the season) whereas low values mean limited epidemics; α_k is related to the slope at the inflection point for trial k, it reflects the speed of epidemic around the peak. Roughly, high values of α mean fast epidemics and low values mean slow epidemics; μ and α are positive parameters; and γ is bounded 0 and 1.

To assess the relative influence of μ, γ and α on the average fitted virus incidence (i.e., $\bar{y} = \frac{1}{55}\sum_{t=1}^{55} y_t$ for a given combination of μ, γ and α), we ran a sensitivity analysis. For this, 50,000 different combinations of the three parameters were randomly drawn within their respective variation ranges (delimited by the extreme values found in the 11 trials, see Results) via a Latin hypersquare sampling method, and sensitivity indices were estimated using Sobol–Saltelli's method [13]. The first-order index of a parameter indicates its main influence on the model output, whereas the total index also includes its interactions with other parameters. Given the negligible influence of parameter α (see Results), the following analyses focused only on $AUDPC_{56}$, μ and γ.

2.5. Computation of Variables Related to Aphid Abundance

The dataset of daily aphid abundance was used to compute, for each of the three main vector species of CABYV (*A. gossypii*, *M. persicae* and *M. euphorbiae*) as well as for the total number of aphids, a wide range of aggregated "aphid" variables tested for their relationship with the virus variables. Firstly, daily abundance was aggregated on periods relative to the planting date, by calculating the sum from time t_1 ($t_1 \in [\![1; 55]\!]$) to time t_2 ($t_2 \in [\![t_1; 55]\!]$), resulting in 1540 different variables for each aphid species. Secondly, daily abundance was aggregated on periods relative to the date of epidemic peak (estimated with parameter μ of the logistic curve), by calculating the sum from time $t_1 = \mu - \Delta t_1$ ($\Delta t_1 \in [\![1; \mu]\!]$) to time $t_2 = \mu - \Delta t_1 + \Delta t_2$ ($\Delta t_2 \in [\![1; 55 - \mu + \Delta t1]\!]$). Depending on the value of μ, this resulted in a maximum of 3025 additional variables.

2.6. Relationship between Aphid and Virus Variables

For each of the three virus variables ($AUDPC_{56}$, μ and γ), a relationship with one or several aphid variables was established in three steps. In a first step, we used the Spearman test with a maximal type-1 error of 1% to identify aphid variables that were significantly correlated to the virus variable under consideration. In a second step, for each remaining aphid variable, we modelled the relationship between the virus variable (dependent variable) and the aphid variable (explanatory variable). For $AUDPC_{56}$, we used the following linear regression (3):

$$z_k = A_0 + A_1.x_k \qquad (3)$$

and for μ and γ, given the shape of data, we used an exponential model (estimated using nonlinear least squares) (4):

$$z_k = B_0 + B_1\left(1 - e^{-B_2.x_k}\right) \qquad (4)$$

with:

z_k—the value of the virus variable (i.e., $AUDPC_{56}$, μ or γ) for trial k ($k \in [\![1; 11]\!]$);

x_k—the value of the aphid variable for trial k;

A_0 and A_1—the parameters of the linear model for $AUPDC_{56}$;

B_0, B_1 and B_2—the parameters of the exponential model for μ and γ, such as $z_k(0) = B_0$ and $z_k(\infty) = B_0 + B_1$.

The mean square error (MSE) was used to evaluate the goodness-of-fit of every model and thus to rank aphid variables according to their potential to explain the virus variable. Finally, in a third step, the model associated with the lowest mean square error was considered as the best candidate to relate aphid variables and the virus variable under consideration. In addition, these best candidates were used to predict values of μ and γ that were themselves used in the logistic equation to rebuild viral epidemic dynamics in each trial.

2.7. Data & Software

Analyses were performed using the R software version 3.5.2 [14]. The sensitivity analysis used the package "sensitivity" version 1.17.1 [15]. Aphid raw data are hosted in a public repository: Data INRAE (Dataverse). Direct URL to data is: https://doi.org/10.15454/NKRWEO.

3. Results

3.1. Virus Epidemics

CABYV was consistently detected in every trial, with epidemic types (based on mean incidence) varying from mild to extreme (Table 2).

Table 2. CABYV epidemics and winged aphid abundances in melon crops in eleven field trials conducted in Avignon between 2010 and 2019. Epidemics are summarized by their area under the disease progress curve calculated over 56 days ($AUDPC_{56}$), mean incidence ($AUDPC_{56}/56$), epidemic category and parameter estimates of the logistic models (μ, γ and α) fitted to cumulative incidences. *Aphis gossypii* (RIS-181), *Myzus persicae* (RIS-322), *Macrosiphum euphorbiae* (RIS-410) and total aphid abundances were monitored with suction traps between 1 and 55 days after planting.

Trial	$AUDPC_{56}$	Mean Incidence (%)	Epidemic Category [a]	μ	γ	α	RIS-181	RIS-322	RIS-410	Total Aphids
M10	3635	65	Extreme	20	1.00	0.141	1693	110	0	3468
M18	2443	44	Severe	21	0.71	0.046	90	5	0	810
M19	697	12	Mild	26	0.24	0.078	76	7	3	841
P11	3540	63	Extreme	21	1.00	0.066	776	72	0	3113
P12	2878	51	Severe	22	0.86	0.051	207	24	0	4004
P13	1330	24	Intermediate	40	0.86	0.045	506	49	0	1772
P14	1617	29	Intermediate	32	0.70	0.037	277	139	0	1382
P15	2834	51	Severe	27	0.99	0.044	407	53	0	2271
V11	2502	45	Severe	32	1.00	0.067	256	379	1	2488
V12	1379	25	Intermediate	36	0.71	0.028	317	118	2	4097
V13	671	12	Mild	51	1.00	0.066	246	46	3	1468

[a] On the basis of their mean incidence, epidemics were categorized as mild (0–20%), intermediate (21–40%), severe (41–60%) or extreme (61–100%) [7].

CABYV disease progress curves were successfully described by the logistic model (Figure 1). Parameter μ (inflection point of the curve indicating the date at which 50% of the maximum incidence is reached) varied between 20 days (M10) and 51 days (V13) (Table 2). Parameter γ indicating the maximum incidence varied between 0.24 (M19) and 1 (M10, P11, V11, V13). Parameter α reflecting the increase rate of disease incidence per day varied between 0.028 (V12) and 0.141 (M10). Taken individually, high values of γ and α, or low values of μ do not necessarily imply severe or extreme epidemics (e.g., $\gamma = 1$ in V13 but the epidemic is mild because α is low and μ is high). There was no correlation among these parameters.

The average virus incidence was mostly influenced by parameters μ and γ, as indicated by their first-order Sobol's sensitivity indices of 0.55 and 0.36, respectively (meaning that 55% and 36% of the variability in average virus incidence can be attributed to the variability in the value of μ and γ, respectively) (Figure 2). The influence of α was negligible, with a total index (which measures the influence of a parameter including its interactions with other parameters) of 0.0013.

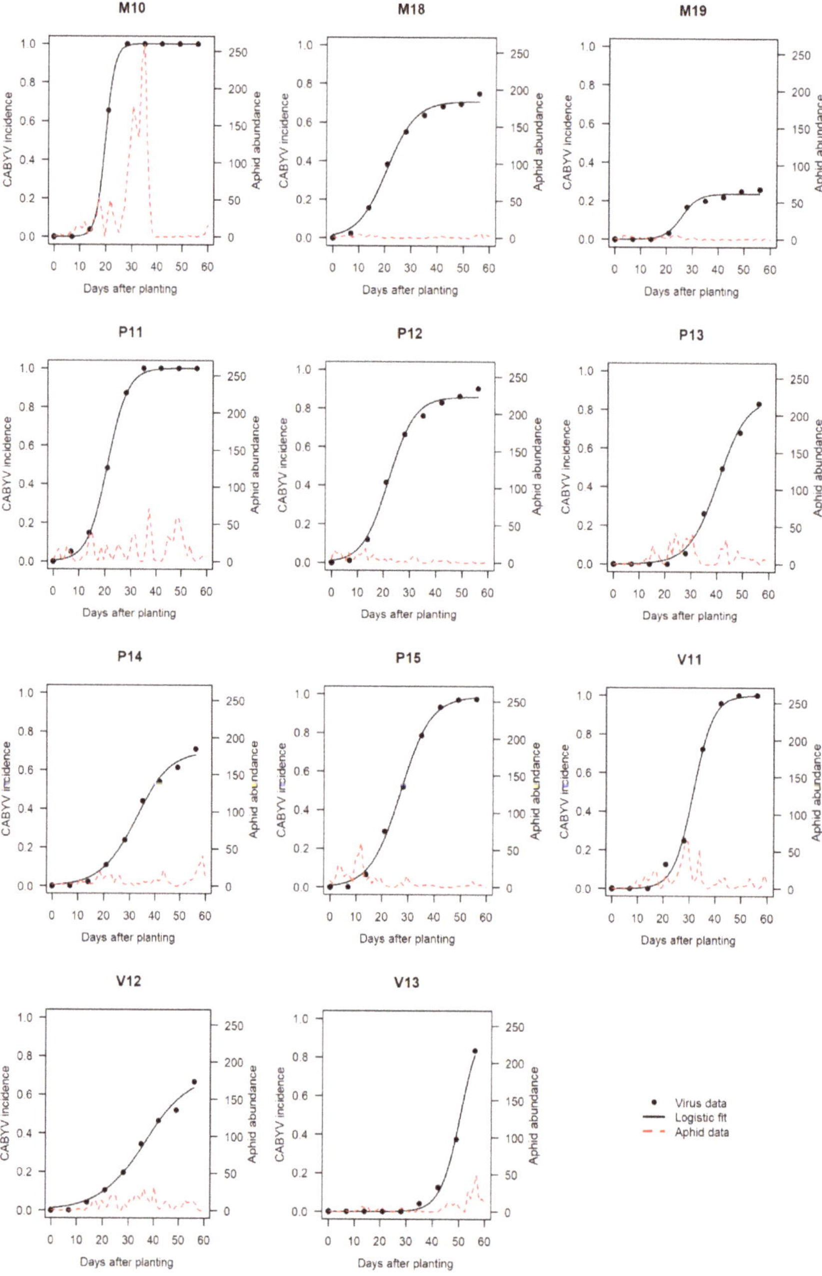

Figure 1. CABYV disease progress and aphid abundance assessed in melon crops in eleven field trials conducted in Avignon between 2010 and 2019. Black dots represent observed cumulative incidences (proportion of infected plants expressed as a ratio). Black solid lines are fitted curves (logistic model). Red dashed lines represent daily abundances of the pool of CABYV aphid vectors (*Aphis gossypii*, *Myzus persicae*, *Macrosiphum euphorbiae*).

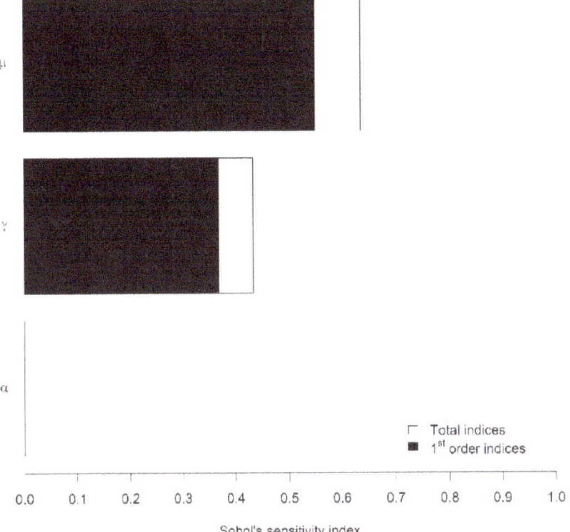

Figure 2. First-order and total Sobol's sensitivity indices of the three parameters of the logistic equation on the average virus incidence. μ is the abscissa of the inflection point (i.e., the date of the epidemic peak); γ is the plateau (i.e., the carrying capacity); α is related to the slope at the inflection point (i.e., the speed of epidemic around the peak). The first-order index indicates the main influence of a parameter, whereas the total index includes its interactions with other parameters.

3.2. Vector Abundances

The three reported CABYV vectors *A. gossypii* (RIS-181), *M. persicae* (RIS-322) and *M. euphorbiae* (RIS-410) represented generally 10% to 30% of the total aphid abundance and could exceptionally reach 52% in M10 (Table 2). *A. gossypii* and *M. persicae* were trapped in every trial. *M. euphorbiae* was present in 4 of the 11 trials and its specific abundance did not exceed three individuals per sampling campaign. With specific abundance representing up to 95% of the total vector abundance, *A. gossypii* was the most abundant vector species in all trials except V11. In V11, *M. persicae* preponderated in catches (60% of the total abundance). As for virus epidemics, patterns of aphid vector dynamics were extremely variable depending on the trials (Figure 1). In some cases, vector activity was more intense at the beginning of the crop (P15 for instance), at mid-crop (M10) or later (V13). *A. gossypii* and *M. persicae* showed dissimilar temporal patterns suggesting a dissimilar host reservoir location and/or dispersal timing (Figure S1).

3.3. Correlations between Virus and Aphid Variables

The large variability in both virus epidemics and aphid abundance dynamics constituted a perfect framework to study the virus–aphid link through a systematic search for correlations between three virus variables ($AUDPC_{56}$, μ and γ; parameter α was not included because its influence on virus incidence was negligible) and more than 9000 aphid variables. These aphid variables were computed by aggregating abundances on periods relative either to the planting date, or to the date of the epidemic peak (i.e., μ, the abscissa of the inflection point of the logistic curve). Depending on the virus variable under consideration, the Spearman test yielded a diverse number of significant correlations with one or several aphid variables. For $AUDPC_{56}$, 113 significant correlations were obtained with abundances of *A. gossypii* or the total aphid population aggregated on periods relative to the planting date (Table S1), and one correlation was obtained with *A. gossypii* abundance aggregated on a period relative to the date of epidemic peak (four consecutive days starting from 11 days before the epidemic peak) (Table S2). Parameter μ was correlated to aphid variables involving either *A. gossypii* or the total aphid population aggregated on periods of 1 to 10 consecutive days within the two first weeks of cropping (Table S1). For parameter γ, 10 significant correlations were obtained with abundances of *A. gossypii* or *M. persicae* aggregated on periods of 1 to 9 consecutive days within the three first weeks of cropping (Table S1, Figure S2A). Ten supplementary significant correlations were found with *A. gossypii* abundances aggregated on periods of 1 to 12 consecutive days before or around the inflection point (Table S2, Figure S2B).

3.4. Selection of the Best Aphid Variables Based on Their Potential to Explain Virus Variables

We used the significant correlations previously identified to build linear models to explain $AUDPC_{56}$ and exponential models to relate μ and γ with aphid variables used as single explanatory variables. Among these models, we selected those associated with the lowest mean square error (MSE). The best linear model to explain the variability of $AUDPC_{56}$ was obtained with the abundance of *A. gossypii* aggregated between D11 and D17 (Figure 3). The variability of parameter μ was best explained by a negative exponential model involving the abundance of *A. gossypii* aggregated between D1 and D10. With regard to parameter γ, the best exponential model involved the abundance of *A. gossypii* aggregated between D12 and D14.

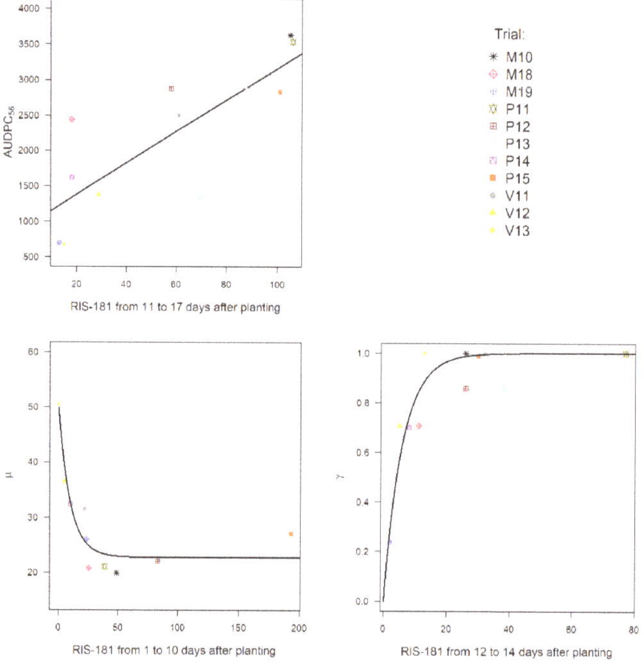

Figure 3. Best models obtained between the virus variables (dependent) and the aphid variables (explanatory). A linear model ($z_k = A_0 + A_1.x_k$) was used for the area under the disease progress curve (AUDPC) and an exponential model ($z_k = B_0 + B_1\left(1 - e^{-B_2.x_k}\right)$) was used for two parameters of the logistic equation (μ, γ).

3.5. Prediction of CABYV Epidemics

The best aphid variables selected at the previous step were used to predict values of parameters μ and γ that, in turn, were used in the logistic equation to rebuild CABYV epidemics in each trial, with α fixed at its mean value 0.061 (Figure 4). The overall shapes of predicted dynamics were in agreement with observed ones and epidemic typology (mild to extreme) was generally maintained.

In 6 of the 11 cases, predictions slightly overestimated the actual CABYV incidence, due to either an overestimation of parameter γ (M18, P12, P13, P14) or an underestimation of parameter μ (P15, V11). In 3 of the 11 cases, predicted and observed CABYV dynamics coincided (M19, P11, V13). In 2 of the 11 cases, predictions slightly underestimated the actual CABYV incidence, due to an overestimation of parameter μ (M10) or underestimation of parameter γ (V12).

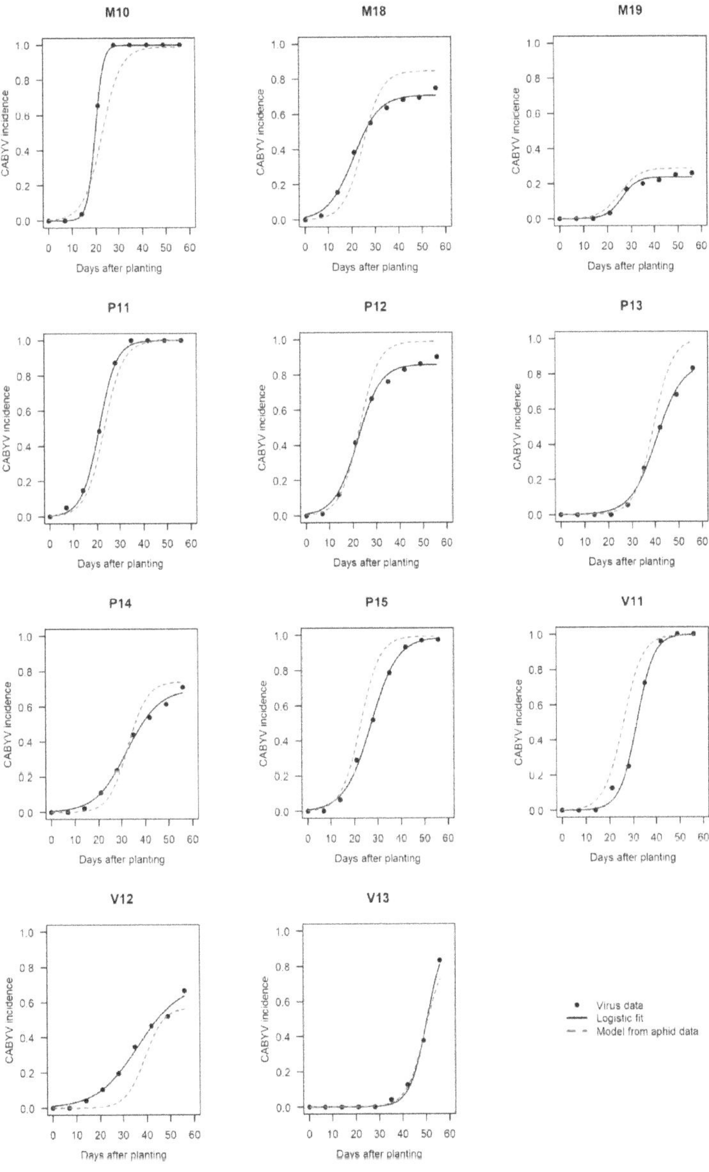

Figure 4. Observed and modelled CABYV epidemic dynamics in melon crops for eleven field trials conducted in Avignon between 2010 and 2019. Black dots represent observed cumulative incidences (proportion of infected plants expressed as a ratio). Black solid lines are fitted curves (logistic model). Red dashed lines represent rebuilt dynamics from the best predictive aphid variables.

4. Discussion

CABYV epidemics were observed in all eleven field experiments conducted between 2010 and 2019 in Avignon, confirming that among the viruses frequently infecting melon crops, namely cucumber mosaic virus (CMV, Cucumovirus, Bromoviridae), watermelon mosaic virus (WMV, Potyvirus, Potyviridae) and zucchini yellow mosaic virus (ZYMV, Potyvirus, Potyviridae), it has become one of the most prevalent. This situation is consistent with recent observations made in the French Mediterranean basin [3] and other countries [4]. When present, other cucurbit viruses do not seem to interfere with CABYV. For example, Schoeny et al. [7] observed a complete decoupling between the progress curves of CABYV, CMV and WMV during the cropping season, suggesting that biotic and/or abiotic factors involved in the epidemiology of these viruses are different. In the present study, CABYV progress over time (expressed as days after planting or fleece removal) was successfully described by the logistic model. This model commonly used to describe the temporal dynamics of plant viruses [16–19] has three parameters (μ, γ, α) with a biological sense (epidemic precocity, carrying capacity, epidemic speed) that were considered as dependent variables in statistical analyses and data mining. Parameters were uncorrelated. In particular, there was no correlation between μ and α, suggesting that early epidemics do not necessarily rise faster than late epidemics contrary to what has been observed in some pathosystems such as virus yellows disease in sugar beet where an increasing host resistance with plant age to feeding aphids is documented [20].

The sensitivity analysis run on the logistic model using randomly drawn combinations of these three parameters revealed that parameters μ (inflection point of the curve) and γ (maximum incidence) had a strong influence on the variability in the average CABYV incidence, whereas parameter α reflecting the disease increase rate per day had a negligible influence on the variability in virus incidence. Parameter μ appeared predominant since its variability could explain 55% of the variability in virus incidence. In our experimental conditions, values of 20–22 days for μ induced systematically severe or extreme epidemics, regardless of the value of γ. With later inflection points, epidemics could be mild to severe depending on γ. Therefore, the earliness of a virus epidemic determines greatly its destiny and consequently its impact on yield. Indeed, the earlier a plant is infected the more yield components are penalized. For CABYV, although not clearly documented, the timing of the virus epidemic compared to the flowering period is likely to be decisive in the fruit development since CABYV is known to induce flower abortion, and consequently a reduction in the number of fruits per plant and an increase in unmarketable over caliber fruits [2].

CABYV being phloem-limited, its acquisition from an infested plant and inoculation to a healthy plant require a phloem-feeding phase by the aphid vector. Contrary to viruses transmitted in a nonpersistent manner through brief intracellular probes into epidermal and/or mesophyll cells by numerous visiting aphids searching for a suitable host, CABYV is transmitted by only a few aphid species (*A. gossypii*, *M. euphorbiae* and *M. persicae*) [5]. Our study focused on this short list of potential vectors.

Winged aphids were monitored daily at the crop height with a non-biased suction trap [12]. Unlike the Rothamsted Insect Survey suction trap monitoring aphid migration flights at a height of 12.2 m above ground [21], our trap sampled winged aphids actually visiting the crop, and possibly transmitting viruses. Among the focused aphid species, *M. euphorbiae* was almost absent from catches, whereas *A. gossypii* and *M. persicae* were present in all trials, with *A. gossypii* being predominant in 10/11 trials. Aphid and virus dynamics were monitored on concomitant periods which facilitated the search for correlations between contemporary events. Although significant correlations were found with aphid variables involving both aphid species as well as total vector abundances, the best correlations involved *A. gossypii*. This is in agreement with the fact that this aphid species is the only aphid species consistently observed feeding and developing colonies on melon crops in France (Boissot, pers. com).

Among the aphid variables highly correlated to virus variables, some appeared as significant explanatory variables. Therefore, the variability of $AUDPC_{56}$ could be explained by the abundance of *A. gossypii* aggregated between D11 and D17 using a simple linear model. The variability of μ and γ

were respectively explained by the abundance of *A. gossypii* aggregated between D1 and D10, and between D12 and D14 using exponential models. It is noteworthy that these two parameters can be predicted as early as two weeks after planting/fleece removal. Thereby, using these predicted parameter values in the logistic equation, it is possible to have an early insight into the probable CABYV dynamic.

This early prediction could permit the implementation of tactical control measures destined to control *A. gossypii* populations. Among possible control measures, the use of insecticides could be optimized by guiding the positioning of treatment in time and space according to the abundance of *A. gossypii* during the two first weeks of the melon crop. Conversely, unnecessary treatments could be avoided if the abundance of *A. gossypii* during this period is low.

A disease forecast prior to planting would be of even greater usefulness. For example, Congdon et al. [22] developed an empirical model to forecast pea seed-borne mosaic virus (PSbMV) incidence in field pea crops using pre-growing season rainfall to calculate an index of aphid abundance which is used in combination with the virus infection level in the sown seed, to provide forecasts before sowing to allow sufficient time to implement control recommendations. Similarly, Steinger et al. [23] observed that the post-harvest incidence of potato virus Y (PVY) in seed potato in Switzerland could be accurately predicted by the cumulative abundance of *Brachycaudus helichrysi* (from first appearance in spring up to mid-June) and that this abundance was positively correlated to the mean daily winter temperature (January–February) indicating that winter conditions could be used as an early warning signal for PVY risk in the current season. Therefore, regarding our pathosystem, forecasting *A. gossypii* spring flying patterns as a function of winter climatic conditions could be worth investigating in order to deploy strategic control measures before planting. For example, whenever available, the selection of a resistant cultivar is an efficient and environmentally friendly way of reducing a disease risk. Concerning the melon crop, *Vat* is a gene conferring resistance to both *A. gossypii* and the viruses it carries [11]. In particular, a five-year field experiment demonstrated that *Vat* had a significant impact on CABYV epidemics with mean incidence reduction exceeding 50% in some trials [7]. Cultural practices such as the use of plastic mulches acting as an aphid repellent [24], floating row covers to prevent viruliferous aphids reaching the crop until the flowering stage or weeding to remove virus reservoirs [8] could complement this genetic resistance. Indeed, CABYV infects cucurbit crops (cucumber, melon, squash and watermelon) but also weed species such as *Capsella bursa-pastoris*, *Lamium amplexicaule* and *Senecio vulgaris*, which may be efficient alternative hosts [1] and more recently, Kassem et al. [6] suggested the importance of the weed species *Ecballium elaterium* as an alternative host and potential virus reservoir. Finally, the implementation of flower strips composed of rigorously selected plant species could also contribute to regulate the populations of aphid vectors by favoring natural enemies [25].

To conclude, our results suggest that the abundance of *A. gossypii* visiting the melon crop during the first fortnight is a good predictor of the CABYV risk, information that could be integrated in a decision support system to improve the efficiency and durability of chemical control. As recently demonstrated by Schoeny et al. [7], *A. gossypii* can also be highly involved in CMV epidemics. Therefore, early flights of *A. gossypii* represent a high virus risk but also a high infestation risk by *A. gossypii* clones capable of developing colonies on melon crops [26]. Therefore, an accurate prediction of this global risk is likely to limit economically unjustified treatments and limit their negative impact on the surrounding environment.

Supplementary Materials: The following materials are available online at http://www.mdpi.com/1999-4915/12/9/911/s1, Figure S1: Daily abundance of CABYV aphid vectors monitored in melons crops with non-biased suction traps in eleven field trials conducted in Avignon between 2010 and 2019. Aphis gossypii (RIS-181) in black solid line; Myzus persicae (RIS-322) in red dashed lines; Macrosiphum euphorbiae (RIS-410) in blue dotted lines, Figure S2: Significant correlations between parameter γ (carrying capacity of the logistic model) and aphid variables calculated: (A) on periods relative to the planting date (i.e., from t1 to t2 days after planting); (B) on periods relative to the date of epidemic peak (i.e., from t1 = μ − Δt1 to t2 = μ − Δt1 + Δt2). Horizontal bars represent the periods over which aphid abundances are aggregated. RIS-181: Aphis gossypii; RIS-322: Myzus persicae; RIS-410: Macrosiphum euphorbiae; total: total aphid population. Correlations are ranked according to the mean square error (MSE) of the corresponding models, Table S1: Significant correlations (with a maximal type-1 error of 1%) between virus variables and aphid variables computed on periods relative to the planting date. Significant correlations involve abundances of Aphis gossypii (RIS-181), Myzus persicae (RIS-322) and total aphid population (total) aggregated between t1 and t2 (in days from planting date). The relationship between the virus variable and the aphid variable was modelled with a linear model ($z_k = A_0 + A_1 \cdot x_k$) for the area under the disease progress curve (AUDPC$_{56}$) and an exponential model ($z_k = B_0 + B_1\left(1 - e^{-B_2 \cdot x_k}\right)$) for the parameters of the logistic equation (μ and γ). For AUDPC$_{56}$, only the 15 correlations having the lowest mean square errors are presented (total of 413), Table S2: Significant correlations (with a maximal type-1 error of 1%) between virus variables and aphid variables computed on periods relative to the date of epidemic peak (μ). Significant correlations involve abundances of *Aphis gossypii* (RIS-181) aggregated between $t_1 = \mu - \Delta t_1$ and $t_2 = \mu - \Delta t_1 + \Delta t_2$ (in days from planting date). The relationship between the virus variable and the aphid variable was modelled with a linear model ($z_k = A_0 + A_1.x_k$) for the area under the disease progress curve (AUDPC$_{56}$) and an exponential model ($z_k = B_0 + B_1\left(1 - e^{-B_2 \cdot x_k}\right)$) for the parameter γ of the logistic equation. No aphid variable was found significantly correlated to μ.

Author Contributions: Conceptualization, A.S.; methodology, A.S. and L.R.; software, L.R.; formal analysis, A.S. and L.R.; investigation, A.S., P.G., G.G., P.M., K.N. and C.W.-S.; writing—original draft preparation, A.S. and L.R.; writing—review and editing, A.S., L.R. and H.L.; supervision, A.S.; project administration, A.S.; funding acquisition, A.S. and H.L. All authors have read and agreed to the published version of the manuscript.

Funding: This study was funded by INRAE (INRA-AAP2010 Gestion durable des résistances-ParcelR), Agence Nationale de la Recherche (ANR-2010-STRA-001-01 VirAphid) and Ministère de l'Agriculture, de l'Alimentation et de l'Environnement (CTPS 25-C-2011-09 and CASDAR 12-1278 Agath).

Acknowledgments: We thank the staff of the INRAE Experimental facilities of the Plant Pathology research unit (IEPV, https://doi.org/10.15454/8DGF-QF70) for their involvement in field experiments; Frédéric Pascal for his technical assistance in suction trap setting-up; Jonathan Gaudin, Antoine Lauvernay and trainees for their help in insect collecting; Philippe Nicot for constructive comments on this manuscript.

Conflicts of Interest: The authors declare no conflict of interest.

References

1. Lecoq, H.; Bourdin, D.; Wipfscheibel, C.; Bon, M.; Lot, H.; Lemaire, O.; Herrbach, E. A new yellowing disease of cucurbits caused by a luteovirus, cucurbit aphid-borne yellows virus. *Plant Pathol.* **1992**, *41*, 749–761. [CrossRef]

2. CABI. Cucurbit aphid-borne yellows virus (original text by Lecoq, H.; Desbiez, C.; and Schoeny, A.). In *Invasive Species Compendium*; CABI: Wallingford, UK, 2020; Available online: http://www.cabi.org/isc/datasheet/110067 (accessed on 17 June 2020).

3. Desbiez, C.; Wipf-Scheibel, C.; Millot, P.; Berthier, K.; Girardot, G.; Gognalons, P.; Hirsch, J.; Moury, B.; Nozeran, K.; Piry, S.; et al. Distribution and evolution of the major viruses infecting cucurbitaceous and solanaceous crops in the French Mediterranean area. *Virus Res.* **2020**, *286*, 198042. [CrossRef] [PubMed]

4. Juarez, M.; Legua, P.; Mengual, C.M.; Kassem, M.A.; Sempere, R.N.; Gomez, P.; Truniger, V.; Aranda, M.A. Relative incidence, spatial distribution and genetic diversity of cucurbit viruses in eastern Spain. *Ann. Appl. Biol.* **2013**, *162*, 362–370. [CrossRef]

5. Lecoq, H.; Desbiez, C. Viruses of Cucurbit Crops in the Mediterranean Region: An Ever-Changing Picture. In *Viruses and Virus Diseases of Vegetables in the Mediterranean Basin*; Elsevier Academic Press Inc.: San Diego, CA, USA, 2012; Volume 84, pp. 67–126.

6. Kassem, M.A.; Juarez, M.; Gomez, P.; Mengual, C.M.; Sempere, R.N.; Plaza, M.; Elena, S.F.; Moreno, A.; Fereres, A.; Aranda, M.A. Genetic Diversity and Potential Vectors and Reservoirs of Cucurbit aphid-borne yellows virus in Southeastern Spain. *Phytopathology* **2013**, *103*, 1188–1197. [CrossRef] [PubMed]

7. Schoeny, A.; Desbiez, C.; Millot, P.; Wipf-Scheibel, C.; Nozeran, K.; Gognalons, P.; Lecoq, H.; Boissot, N. Impact of Vat resistance in melon on viral epidemics and genetic structure of virus populations. *Virus Res.* **2017**, *241*, 105–115. [CrossRef] [PubMed]

8. Lecoq, H.; Katis, N. Control of Cucurbit Viruses. In *Control of Plant Virus Diseases—Seed-Propagated Crops*; Elsevier Academic Press Inc.: San Diego, CA, USA, 2014; Volume 90, pp. 255–296.

9. Dogimont, C.; Bussemakers, A.; Martin, J.; Slama, S.; Lecoq, H.; Pitrat, M. Two complementary recessive genes conferring resistance to Cucurbit Aphid Borne Yellows Luteovirus in an Indian melon line (cucumis melo L.). *Euphytica* **1997**, *96*, 391–395. [CrossRef]

10. Kassem, M.A.; Gosalvez, B.; Garzo, E.; Fereres, A.; Gómez-Guillamón, M.L.; Aranda, M.A. Resistance to Cucurbit aphid-borne yellows virus in Melon Accession TGR-1551. *Phytopathology* **2015**, *105*, 1389–1396. [CrossRef] [PubMed]

11. Boissot, N.; Schoeny, A.; Vanlerberghe-Masutti, F. Vat, an Amazing Gene Conferring Resistance to Aphids and Viruses They Carry: From Molecular Structure to Field Effects. *Front. Plant Sci.* **2016**, *7*, 1420. [CrossRef] [PubMed]

12. Schoeny, A.; Gognalons, P. Data on winged insect dynamics in melon crops in southeastern France. *Data Brief* **2020**, *29*, 105132. [CrossRef] [PubMed]

13. Saltelli, A.; Annoni, P.; Azzini, I.; Campolongo, F.; Ratto, M.; Tarantola, S. Variance based sensitivity analysis of model output. Design and estimator for the total sensitivity index. *Comput. Phys. Commun.* **2010**, *181*, 259–270. [CrossRef]

14. Team, R.C. *R: A Language and Environment for Statistical Computing*; R Foundation for Statistical Computing: Vienna, Austria, 2012.

15. Pujol, G.; Iooss, B.; Janon, A.; Khalid, B.; Da Veiga, S.; Delage, T.; Fruth, J.; Gilquin, L.; Guillaume, J.; Le Gratiet, L.; et al. Sensitivity: Global Sensitivity Analysis of Model Outputs. R Package Version 1.11.0. 2017. Available online: https://CRAN.R-project.org/package=sensitivity (accessed on 2 March 2020).

16. Batista, L.; Velazquez, K.; Estevez, I.; Pena, I.; Lopez, D.; Reyes, M.L.; Rodriguez, D.; Laranjeira, F.F. Spatiotemporal dynamics of Citrus tristeza virus in Cuba. *Plant Pathol.* **2008**, *57*, 427–437. [CrossRef]

17. Moreira, A.S.; Bergamin, A.; Rezende, J.A.M. Comparative Epidemiology of Three Virus Diseases on Zucchini Squash. *J. Phytopathol.* **2015**, *163*, 386–394. [CrossRef]

18. Nutter, F.W.J. Quantifying the temporal dynamics of plant virus epidemics: A review. *Crop Prot.* **1997**, *16*, 603–618. [CrossRef]

19. Salvalaggio, A.E.; Lopez Lambertini, P.M.; Cendoya, G.; Huarte, M.A. Temporal and spatial dynamics of Tomato spotted wilt virus and its vector in a potato crop in Argentina. *Ann. Appl. Biol.* **2017**, *171*, 5–14. [CrossRef]

20. Werker, A.R.; Dewar, A.; Harrington, R. Modelling the incidence of virus yellows in sugar beet in the UK in relation to numbers of migrating Myzus persicae. *J. Appl. Ecol.* **1998**, *35*, 811–818. [CrossRef]

21. Macaulay, E.D.M.; Tatchell, G.M.; Taylor, L.R. The Rothamsted Insect Survey '12-metre' suction trap. *Bull. Entomol. Res.* **1988**, *78*, 121–128. [CrossRef]

22. Congdon, B.S.; Coutts, B.A.; Jones, R.A.C.; Renton, M. Forecasting model for Pea seed-borne mosaic virus epidemics in field pea crops in a Mediterranean-type environment. *Virus Res.* **2017**, *241*, 163–171. [CrossRef] [PubMed]

23. Steinger, T.; Goy, G.; Gilliand, H.; Hebeisen, T.; Derron, J. Forecasting virus disease in seed potatoes using flight activity data of aphid vectors. *Ann. Appl. Biol.* **2015**, *166*, 410–419. [CrossRef]

24. Lecoq, H.; Pitrat, M. Effects of resistance on the epidemiology of virus diseases of cucurbits. In *Proceedings of Cucurbitaceae 89: Evaluation and Enhancement of Cucurbit Germplasm*; USDA Vegetable Lab.: Charleston, SC, USA, 1989; pp. 404–408.

25. Schoeny, A.; Lauvernay, A.; Lambion, J.; Mazzia, C.; Capowiez, Y. The beauties and the bugs: A scenario for designing flower strips adapted to aphid management in melon crops. *Biol. Control* **2019**, *136*, 103986. [CrossRef]

26. Thomas, S.; Vanlerberghe-Masutti, F.; Mistral, P.; Loiseau, A.; Boissot, N. Insight into the durability of plant resistance to aphids from a demo-genetic study of Aphis gossypii in melon crops. *Evol. Appl.* **2016**, *9*, 756–768. [CrossRef] [PubMed]

Article

RNAseq Analysis of Rhizomania-Infected Sugar Beet Provides the First Genome Sequence of Beet Necrotic Yellow Vein Virus from the USA and Identifies a Novel Alphanecrovirus and Putative Satellite Viruses

John J. Weiland [1], Roshan Sharma Poudel [2,†], Alyssa Flobinus [1,2], David E. Cook [3], Gary A. Secor [2] and Melvin D. Bolton [1,2,*]

[1] United States Department of Agriculture, Agricultural Research Service, Northern Crop Science Laboratory, Fargo, ND 58102, USA; John.Weiland@usda.gov (J.J.W.); alyssa.flobinus@ndsu.edu (A.F.)

[2] Department of Plant Pathology, North Dakota State University, Fargo, ND 58108, USA; roshan.sharmapoudel@ndsu.edu (R.S.P.); Gary.Secor@ndsu.edu (G.A.S.)

[3] Department of Plant Pathology, Kansas State University, Manhattan, KS 66506, USA; decook@ksu.edu

* Correspondence: melvin.bolton@usda.gov

† Current address: United States Department of Agriculture, Agricultural Research Service, Northern Crop Science Laboratory, Fargo, ND 58102, USA.

Received: 14 May 2020; Accepted: 9 June 2020; Published: 10 June 2020

Abstract: "Rhizomania" of sugar beet is a soilborne disease complex comprised of beet necrotic yellow vein virus (BNYVV) and its plasmodiophorid vector, *Polymyxa betae*. Although BNYVV is considered the causal agent of rhizomania, additional viruses frequently accompany BNYVV in diseased roots. In an effort to better understand the virus cohort present in sugar beet roots exhibiting rhizomania disease symptoms, five independent RNA samples prepared from diseased beet seedlings reared in a greenhouse or from field-grown adult sugar beet plants and enriched for virus particles were subjected to RNAseq. In all but a healthy control sample, the technique was successful at identifying BNYVV and provided sequence reads of sufficient quantity and overlap to assemble >98% of the published genome of the virus. Utilizing the derived consensus sequence of BNYVV, infectious RNA was produced from cDNA clones of RNAs 1 and 2. The approach also enabled the detection of beet soilborne mosaic virus (BSBMV), beet soilborne virus (BSBV), beet black scorch virus (BBSV), and beet virus Q (BVQ), with near-complete genome assembly afforded to BSBMV and BBSV. In one field sample, a novel virus sequence of 3682 nt was assembled with significant sequence similarity and open reading frame (ORF) organization to members within the subgenus *Alphanecrovirus* (genus *Necrovirus*; family *Tombusviridae*). Construction of a DNA clone based on this sequence led to the production of the novel RNA genome in vitro that was capable of inducing local lesion formation on leaves of *Chenopodium quinoa*. Additionally, two previously unreported satellite viruses were revealed in the study; one possessing weak similarity to satellite maize white line mosaic virus and a second possessing moderate similarity to satellite tobacco necrosis virus C. Taken together, the approach provides an efficient pipeline to characterize variation in the BNYVV genome and to document the presence of other viruses potentially associated with disease severity or the ability to overcome resistance genes used for sugar beet rhizomania disease management.

Keywords: sugar beet; rhizomania; RNAseq; virus; necrovirus; helper virus

1. Introduction

The increasing globalization of food and other commodities has resulted in greater exposure of crops to historical and emerging pests and diseases, including viruses [1,2]. Within a pathosystem, novel pathogen variants acting alone or in combination may arise to compromise existing resistance in the crop [3–5]. Because of the unanticipated complexities wrought by such phenomena in disease development and crop protection, accurate, rapid and, increasingly, naïve diagnostics and analyses are required to keep pace with the accelerating diversity in pests and pathogens.

Sugar beet contributes nearly half of the sucrose produced and consumed within the United States of America (USA) [6]. "Rhizomania" is considered to be the most devastating root disease of sugar beet globally and is managed principally through genetic resistance in the host [7,8]. The disease is caused by the beet necrotic yellow vein virus (BNYVV; family *Benyviridae*, genus *Benyvirus*). Like most plant viruses, BNYVV is a positive-sense RNA virus. The genome of BNYVV is divided among 4 or 5 segments (designated RNAs 1 to 5), the roles of which, during infection, have been the source of intense investigation over the past 30 years [9–12]. Thus, genes, as well as non-coding regions, have been assigned functions in virus replication, encapsidation, intra- as well as inter-plant movement, silencing suppression, and pathogen aggressiveness [10,13]. Recent studies on the latter of these functions have led to the discovery of a highly mutable region in the *p25* gene on RNA 3, implicated in the breaking of monogenic resistance provided by the Rz1 gene [5,14–17]. The Rz1 gene was discovered in sugar beet germplasm in the 1980s and, even with the breaking of this resistance by strains of BNYVV worldwide, it remains the most widely used gene for the management of this disease [6]. Nevertheless, many questions remain regarding the full nature of resistance-breaking by BNYVV in sugar beet, the extent to which the plasmodiophorid vector of BNYVV, *Polymyxa betae*, plays a role in disease aggressiveness, and the degree to which other known, and possibly unknown, viruses in the sugar beet root infection court positively or negatively impact disease.

Rapid and accurate genetic analysis of known and potential pathogens is a fundamental goal of disease management programs as the genetic constitution of a pathogen population may be crucial in determining their disease-causing potential. The advent of "next-generation sequencing" (NGS) technologies for DNA and RNA analysis, combined with the increasing power of computational platforms, has removed a long-standing roadblock in rapid, cost-effective, population-scale sequence generation and genome examination [18,19]. In cases where RNA transcripts or RNA viruses are the targets of an investigation, RNAseq has become a widely used approach to obtain sequence representation of all RNA in a sample, with sample fractionation offering a way to bias the RNA population toward desired targets [20,21]. An added power to RNAseq is that it is "naïve", remaining relatively unbiased with respect to the molecules contributing to the final sequence assembly. Thus, RNAseq offers a powerful tool for generating sequenced transcriptomes from multi- and unicellular organisms and for producing RNA "viromes"–the global representation within a sample of all RNA viruses present [22].

In the present study, samples from diseased sugar beet seedlings and adult plants were enriched for viruses and subjected to RNAseq analyses. Reads were assembled to produce genomic segments, or in some cases, open reading frames (ORFs), which were used to identify potential relatedness of the viral genomes obtained to existing viral sequences in public databases. Based on the derived BNYVV sequence of RNAs 1 and 2 and that of a previously uncharacterized necrovirus, infectious cDNA clones were developed, validating the integrity of the approach. With the additional discovery of potentially novel satellite viruses, the results confirm the usefulness of the method in assessing the spectrum of viruses present in sugar beet plants exhibiting rhizomania disease.

2. Materials and Methods

2.1. Sample Collection

Three soil samples (S1–S3) and one root sample (S6) were obtained from the sugar beet production areas of the Red River Valley of North Dakota and Minnesota and southern Minnesota by agriculturalists of the Min-Dak Farmers Cooperative (Wahpeton, ND, USA) and Southern Minnesota Beet Sugar Cooperative (Renville, MN, USA) in 2018 (Table S1). Two soil samples (S4 and S5) were also received from sugar beet production areas in Idaho by agriculturalists of The Amalgamated Sugar Co. (Boise, ID, USA) in 2017 (Table S1).

2.2. Plant Growth and Virus Recovery

To recover BNYVV from infested soil samples, we followed the methodology described by Weiland et al. [17]. Briefly, rhizomania-susceptible sugar beet seeds of the cultivar SLC4-K583-G1C (SES VanderHave; Tienen, Belgium) were sown in 250 mL pots containing one-part infested soil to one-part sterile sand. As a negative control, seeds were planted into sterile sand mixed 1:1 with Sunshine Mix #1 (Sungro Horticulture; Agawam, MA, USA). Slow-release fertilizer (Multicote; Sungro Horticulture; Agawam, MA, USA) was added following the manufacturer's instructions. Plants were grown in a greenhouse under standardized conditions at 24 °C (day)/18 °C (night), with 8 h of supplemental light per day. Water was added directly as needed. Six weeks after planting in infested soil, plants were harvested, and a root sample consisting of 5 to 7 seedlings was taken from each pot. Roots were washed under running tap water, tamped dry on paper toweling, and stored on ice in plastic bags in preparation for virus extraction. Freezing of samples was avoided, which can compromise the integrity of the viral RNA upon sample thawing. ELISA to detect BNYVV was performed on parallel samples in order to determine the best replicates to use for virus extraction. For Sample 6, comprised of single mature, field-grown sugar beet root, the hairy roots characteristic of the disease were washed thoroughly in tap water, excised from the root surface using care to include a portion of the necrotic veinal tissue in the sample and processed for virus enrichment as described below.

2.3. Virus Enrichment

Efforts were made to enrich for virus particles using standard PEG precipitation of crude extracts [23]. Briefly, fresh root tissue was ground with a mortar and pestle in cold 0.1 M $NaPO_4$ pH 5.2 (3 mL buffer per g fresh weight of tissue). Each homogenized sample was transferred as 1.0 mL aliquots to several 1.5 mL microcentrifuge tubes, each containing 0.4 mL of chloroform, and the contents mixed well and then centrifuged at 16,000× *g* at 4 °C for 18 min. Supernatants were transferred to new tubes, adjusted to 1% NaCl and 8% PEG 8000, and incubated on ice for 10 min. Following a centrifugation step for 18 min at 16,000× *g* at 4 °C, the supernatant was removed and discarded. The whitish pellet was resuspended in 0.4 mL of cold 0.1 M $NaPO_4$ pH 5.2 and the sample extracted with 0.1 mL of chloroform. The aqueous phase from the extraction was precipitated once again by the addition of NaCl to 1% and PEG 8000 to 8%, and samples were incubated on ice overnight. The following day, the samples were centrifuged at 4 °C for 18 min at 16,000× *g*. Care was taken to remove to the extent possible all PEG supernatant from the small pellet. RNA was extracted from the pellet, as previously described in Weiland and Edwards [24]. The integrity of the recovered RNA was determined using agarose gel electrophoresis, and the RNA quantified using spectrophotometry. Shipment of samples to commercial producers of RNAseq data followed the instructions of the contractor.

2.4. RNAseq Analyses

RNAseq libraries (150 bp insert size) were prepared and sequenced (pair-end 100 bp) by BGI Americas (Cambridge, MA, USA) or Admera (South Plainfield, NJ, USA) using the Illumina Highseq 4000 sequencing platform. Customized bioinformatics was also provided by each company. Briefly, low-quality reads and adaptor sequences were removed. For each sample, short reads were de novo assembled with different k-mer sizes in parallel. Reads were subsequently mapped back to the assembled contigs for validation. The best assembly was chosen based on contig N50 and mapping rate. Standard BLAST queries were used to verify or postulate the identification of novel viruses discovered in this work and for confirming the accuracy and completeness of genome assemblies.

To identify known sugar beet viruses, high-quality filtered reads from each sample were mapped to reference genome sequences of BNYVV (GeneBank assembly accession: GCF_000854885.1), BSBMV (GeneBank assembly accession: GCA_002867265.1) and BBSV (GeneBank assembly accession: GCF_000855285.1). Mapping was done using default parameters (except length fraction = 0.8) in CLC genomics workbench v 8.0 (CLC Bio, Qiagen, Germantown, MD, USA). Any reads (paired- and single-end reads) mapped to a given viral genome were extracted for de novo assembly in CLC genomics using default parameters (word size = automatics; bubble size = automatic). A near-complete genome assembly of BNYVV, BSBMV, and BSBV was obtained from the majority of the samples. Assembly of sequences from Sample 3 also suggested the presence of a novel virus with sequence characteristics of plant Alphanecroviruses and potential satellite viruses.

2.5. Sequence Analysis of BNYVV Strains

The de novo generated BNYVV RNA sequences were BLASTn (https://blast.ncbi.nlm.nih.gov) searched against publicly available BNYVV sequences to inspect the percentage identity (homology) between them. Nucleic acid consensus sequences for BNYVV isolates collected as part of this work were designated according to sample number. Assembled sequences of RNAs 1 to 4 of BNYVV were used to infer ORF locations. The analysis of sequence relatedness between RNAs 1 and 2 from this study (GenBank accessions MT227164 and MT227165, respectively) and those in GenBank was performed using BLAST. Due to the larger number and wider geographic distribution of available sequences for BNYVV RNAs 3 and 4, genes *p25* and *p31* (from RNA3 and RNA4, respectively; GenBank accessions MT372831-MT372842) were analyzed. Multiple sequence alignment of *p25* sequences from 52 strains and for *p31* from 46 strains were carried out using MUSCLE v3.8.31 (https://www.ncbi.nlm.nih.gov/pmc/articles/PMC390337/). The phylogenetic relationships were inferred using Randomized Axelerated Maximum Likelihood (RAxML) v8.2.9 (https://www.ncbi.nlm.nih.gov/pmc/articles/PMC3998144/). Our RAxML analysis utilized rapid bootstrap analysis to search for the best-scoring ML tree with the number of bootstrap iterations determined at runtime using the extended majority-rule consensus tree criterion (i.e., bootstopping), and the "GTRGAMMA" model of nucleotide substitutions. The best-scoring ML tree for each gene was visualized in FigTree v1.4.4 (https://github.com/rambaut/figtree). Orthologous genes from BSBMV were tested for inclusion in the analysis to serve as an outgroup, but the sequences were too divergent, a similar conclusion as previously reported [25]. As such, we employed midpoint rooting in FigTree.

2.6. Construction and Inoculation of BNYVV RNA1 and 2 Infectious Clones

Based on the data obtained in the present work from RNAseq and conserved sequences at the 5′- and 3′-termini in BNYVV genomes from across the globe (Table 2), clones of RNA 1 and 2 were designed for construction. For BNYVV RNA 1 cDNA, four synthetic overlapping fragments were generated by Genewiz (South Plainfield, NJ, USA) and delivered as discrete fragments in vector pUC57, which were used as a template to produce four PCR amplicons (primer sequences in Table S2). *Bam*HI-T7 promoter (25 nt) and polyA$_{60}$-*Hind*III-*Bam*HI (72 nt) sequences were incorporated on the 5′ end of the first fragment and the 3′ end of the fourth fragment, respectively, by PCR. Subsequently, these amplicons were used as templates in overlap extension PCR (Table S2) to generate the full-length cDNA of RNA 1. The entire sequence of the RNA1 cDNA (6747 nt genome) was cloned into the pNEB193 vector (New England Biolabs, Waltham, MA, USA) using *Bam*HI restrictions sites. The full-length BNYVV RNA 2 cDNA clone (4610 nt genome) was synthesized by Genewiz and delivered in a pSMART-BAC (Lucigen Inc., Middleton, WI, USA) vector flanked by two *Bam*HI restriction sites. The RNA 1 and RNA 2 clones were linearized by *Bam*HI and *Hind*III restriction enzymes, respectively, and used to produce capped, polyadenlylated RNA using methods previously described [24]. Inoculation of the transcript RNA to leaves of *Chenopodium quinoa* followed the procedure of Petty et al. [26], and Western blotting and detection of BNYVV-infected leaves using an anti-BNYVV antibody (Agdia Inc., Elkhart, IN, USA) were performed according to Weiland and Edwards [24].

2.7. Putative Alphanecrovirus and Satellite Virus Sequence Validation and Characterization

Primers for cDNA synthesis and DNA amplification and sequencing were designed based upon sequences produced through RNAseq and from sequence accessions in public sequence databases. Reverse transcription and polymerase chain reaction (RT-PCR) conditions, using primers indicated below, were as outlined in Edwards et al. [27]. Amplification of putative satellite sequences encoding the predicted coat protein was done using primers MDB-1867 and MDB-1868 (Table S2). Primers MDB-2100 and MDB-2101 (Table S2) were employed to generate a single amplicon from the putative novel Alphanecrovirus in sample S3. The amplicon sequence originates in the 3′-end of the p52 ORF, spans the p8 and p6 ORFs, and encompasses the entire predicted P30 CP gene (Figure 4). The sequence of the P23 ORF subsequently was found within the raw sequence reads extending the assembled sequence towards the 5′-end of the genome. Finally, for both the novel Alphanecrovirus and the satellite virus, the SMART RACE kit (TaKaRa Bio Inc., Mountain View, CA, USA) was employed to capture and characterize 5′- and 3′-end sequences.

2.8. Construction and Inoculation of Novel Alphanecrovirus Infectious Clones

Full-length clone construction for the novel Alphanecrovirus was initiated by generation of a genome-length amplicon using primers MDB-2460 and MDB-2462 (Table S2) in which the first 17 nt of MDB-2460 comprise the phage T7 RNA polymerase promoter. The amplicon was blunt-cloned into pMiniT 2.0 (New England Biolabs, Waltham, MA, USA). Two clones (pBvANV#7 and pBvANV#10) were linearized with *Eco* R1 restriction enzyme and transcribed in vitro, as described previously, for the generation of infectious RNA of the Betanecrovirus BBSV [28]. Inoculation of expanded leaves of healthy *C. quinoa* with the synthetic RNA derived from clones pBvANV #7 and 10 also followed the methods of Weiland et al. [28]. ELISA analysis of protein extracts prepared from diseased and healthy *C. quinoa* leaves employed the same methods as noted above for the detection of BNYVV CP but using a TNV-A detection kit (Loewe Diagnostics, Sauerlach, Germany).

3. Results

3.1. Virus Enrichment and RNAseq Analysis

The use of polyethylene glycol in virus precipitation has been employed in virus purification for decades [23] and provided the basis for the enrichment in the present study. An example of the efficacy in the enrichment of viral RNA over cellular nucleic acids is shown in Figure 1 and Figure 5, where removal of the bulk of rRNA and genomic DNA is evident. Additionally, this is also evident in Figure 5, where a satellite virus genome is clearly the most abundant product of the enrichment scheme. Approximately 20 to 40 ng of prepared RNA sufficed to produce libraries for each sample capable of yielding reads for the assembly of multiple virus genomes present in the samples.

Approximately 60,000,000 RNAseq reads were obtained per sample (Table 1). Using BLAST alignments to assign reads to known virus sequences in the NCBI database, numerous confirmed (Table 1) and potential sugar beet viruses were detected within the samples (see Figure S1 for maps of main viruses observed). The viruses BNYVV, BSBMV [29], and BBSV [28] are known to produce disease symptoms on infected sugar beet plants, whereas plants infected by BSBV [30] and BVQ [31] are relatively asymptomatic. BBSV was first reported in the USA in 2006 in Colorado [32], the present work documenting a more western distribution of this virus in the USA than previously known (near-complete and partial genome sequence found in Samples 4 and 5, respectively, Gooding County, ID, USA). The proportion of the genome able to be assembled for each of the viruses varied between viruses within a sample but was over 92% for all BNYVV RNAs (Table 1). The viruses BNYVV and BSBMV were generally associated with the highest proportion of viral sequence reads. This is consistent with the fact that the study targeted samples with the highest probability of inducing Rhizomania disease. The genome sequence of BSBMV was nearly identical to that reported by Lee et al. [33], with less than 0.01% nucleotide differences observed. Sample S6 differed from all other samples in being obtained from field-grown sugar beets in southern Minnesota. The virus complement was not markedly different from those obtained through baiting of the viruses from soil in a greenhouse setting, even as the number of reads of BNYVV predominated in the sample.

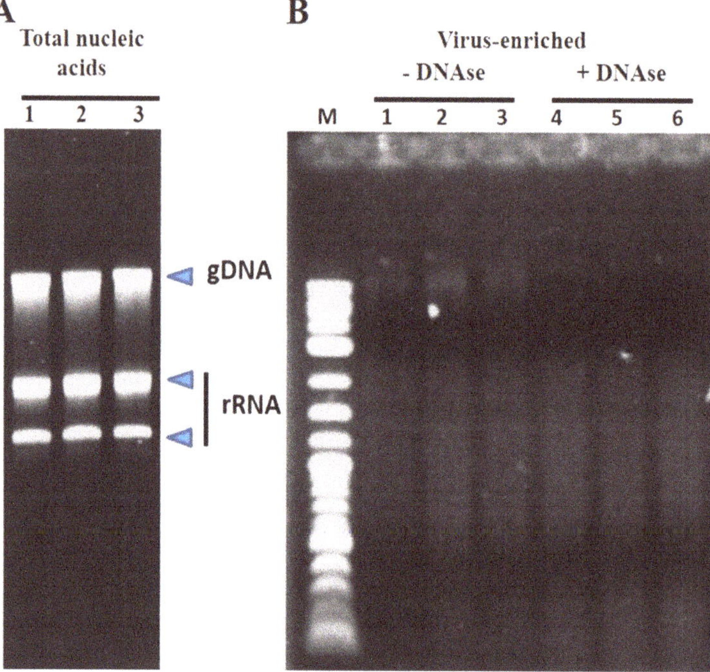

Figure 1. Agarose gel electrophoresis comparing total RNA prepared from sugar beet roots versus using virus-enriched RNA. (**A**) Typical preparation of total nucleic acids from roots of 6-week-old sugar beet plants is dominated by genomic DNA and ribosomal RNA. (**B**) Virus-enrichment removes the bulk of the cellular RNA, although additional DNase treatment (compare -DNase and +DNase lanes) is required to remove residual genomic DNA in preparing the sample for RNAseq. A DNA size standard (M) is included to monitor the approximate size of the RNA population.

Table 1. RNAseq reads mapped to known sugar beet viruses.

Beet Necrotic Yellow Vein Virus (GCF_000854885.1; 15,914 nt) [a]													
RNA 1 (NC_003514.1; 6746 nt) [b]				RNA 2 (NC_003515.1; 4609 nt)				RNA 3 (NC_003516.1; 1774 nt)				RNA 4 (NC_003517.1; 1465 nt)					
#	Start	End	Reads Mapped	Start	End	Reads Mapped	Start	End	Reads Mapped	Start	End	Reads Mapped	Total Reads Mapped	Total Reads
S1	12	6735	46,701	14	4594	82,921	17	1753	36,298	12	1465	37,957	203,877	66,078,080
S2	12	6734	121,686	16	4595	240,749	14	1760	71,911	14	1451	123,978	558,325	69,414,580
S3	10	6734	97,617	11	4595	235,333	17	1758	52,887	14	1448	93,530	479,367	71,362,360
S4	2	6724	15,646	18	4609	21,455	12	1766	12,608	14	1465	5373	55,082	62,246,430
S5	6	6727	8369	19	4596	15,041	22	1761	6411	14	1465	2840	32,661	59,595,934
S6	5	6744	26,060	16	4609	40,054	1	1744	25,411	14	1465	19,449	110,974	73,945,544

Beet Soilborne Mosaic Virus (GCA_002867265.1; 14,744 nt)													
RNA 1 (JF513082.1; 6679 nt)				RNA 2 (JF513083.1; 4615 nt)				RNA 3 (EU410955.1; 1720 nt)				RNA 4 (FJ424610.2; 1730 nt)					
#	Start	End	Reads Mapped	Start	End	Reads Mapped	Start	End	Reads Mapped	Start	End	Reads Mapped	Total Reads Mapped	Total Reads
S1	12	6671	140,396	16	4597	424,451	1	1720	245,901	1	1714	67,780	878,528	66,078,080
S2	12	6671	519,420	16	4595	1,428,890	15	1720	827,273	13	1721	131,926	2,902,509	69,414,580
S3	12	6668	153,627	16	4599	677,253	13	1720	217,594	13	1717	162,226	1,210,700	71,362,360
S4	7	6676	8696	16	4615	16,635	1	1720	16,177	13	1730	2210	43,718	62,246,430
S5	6	6679	83,687	17	4615	157,895	13	1720	133,978	1	1730	56,128	431,688	59,595,934
S6	6	6646	738	33	4598	853	16	1702	136	24	1714	24	1751	73,945,544

[a] Indicates NCBI Accession used as reference and size of genome in nucleotides (nt). Note that 15,914 nt is the size of the BNYVV genome with RNA 5, which is lacking in the United States (USA) strains of the virus; [b] Indicates NCBI Accession used as reference and size of genome in nucleotides (nt).

3.2. Taxonomic Grouping of US BNYVV Isolates and Biological Validation of Sequences

Analysis of the BNYVV genomes obtained in the study indicated low-level variation in the sequence of RNAs 3 and 4 from different sampling locations, but no significant variation between sampling sites for RNAs 1 and 2. The most variable region of the BNYVV genome worldwide is the noted "tetrad region" encoded on RNA3 between nucleotides 645 and 656 (E12 strain, NCBI Accession #EU330455.1; [5]). The most common tetrad amino acids (AAs) observed in the samples of the present study included VCHG, ACHG, VLHG, and TLHG, the first three of which have a prior association with the breaking of Rz1-gene-based resistance in sugar beet [5,14,17]. All sequences were used to confirm the placement of the BNYVV-US within phylogenetic groupings (Figure 2) previously reported by Chiba et al. [23], Schirmer et al. [34], and Zhuo et al. [35]. Through the ranking of similarity with previous sequences already used in phylogenetic group formation (Table 2; Figure 2), the lineage of the US isolates obtained based on sequence analysis was in broad agreement with the results of these prior analyses, indicating that these viruses are of the A-type designation. Additionally, the sample sequences provided evidence of the continued absence of RNA-5 from the US, as has been noted previously [6].

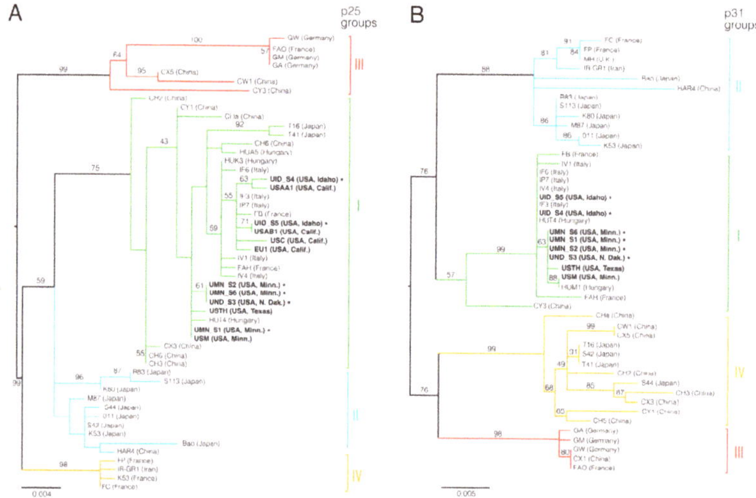

Figure 2. Relatedness of USA BNYVV to global isolates based on *p25* and *p31* gene sequences. Phylogenetic trees were inferred using Randomized Axelerated Maximum Likelihood (RAxML), using runtime calculated bootstrapping. (**A**) Coding sequence of p25 from 52 BNYVV strains and (**B**) coding sequence of p31 from 46 strains, coded on RNA3 and RNA4 of the BNYVV genome, respectively. Each tree was midpoint rooted, and ML support values are displayed for each branch with greater than 50% support. The four resulting groups in each tree were colored and indicated with solid bars shown to the right of each tree. The strain name is shown for each isolate along with the reported country of isolation in parentheses along with the state for US isolates. All US isolates are in bold. Sequences collected from this study are denoted with *.

3.3. Development of Infectious Clones of BNYVV RNAs 1 and 2

Given that RNAseq produces a consensus sequence from an RNA virus population that may exhibit underlying variation and that RNAs 1 and 2, nevertheless, were highly invariant across samples, we reasoned that DNA clones synthesized based on the obtained sequence would have a high probability of being infectious. However, the obtained sequences did not include the terminal nucleotides represented in previously confirmed sequences of BNYVV (Table 1). Alignment of RNA 1 and 2 sequences presented in Table 2 confirmed that the terminal sequences appeared invariant in BNYVV across continents. Consequently, we incorporated these nucleotides not present in the RNAseq data in the terminal primer sequences towards the production of infectious BNYVV RNA 1 and RNA 2 clones. With the historical difficulty in producing and maintaining clones of RNA 1 and RNA 2 in multiple laboratories across the globe, we elected to construct these in a BAC vector using a commercial gene synthesis service. Success in this approach was achieved with RNA 2, but RNA 1 proved intractable for cloning by this method. Instead, an RNA 1 clone was obtained within our research laboratory in which the insert cDNA copy was (a) produced as a full-length 6.7 kb amplicon from source materials, (b) identical in sequence to that submitted to the commercial service for synthesis, and (c) successfully maintained and amplified in pNEB193, a high-copy number plasmid vector. Linearization of both the RNA-1-possessing plasmid clone and the RNA-2-possessing BAC clone permitted the synthesis in vitro of capped transcripts representing both genome components. Infectivity of the capped, poly-adenylated transcripts produced abundant lesions on inoculated *Chenopodium quinoa* plants, the cause of which being due to BNYVV infection was confirmed by Western blot analysis (Figure 3). A more complete description of the clones and their use in BNYVV variant analysis is in progress in our laboratory (Flobinus et al., in preparation).

Table 2. Similarity between the US BNYVV consensus sequence for RNA 1 and RNA 2 and cognate BNYVV RNAs from outside the USA.

Country of Origin	BNYVV RNA 1					BNYVV RNA 2					
	Accession #[a]	Identity (%)	Aligned	Mismatch	Gaps	Accession #	Identity (%)	Aligned	Mismatch	Gaps	Type[b]
Brazil	MH106726.1	99.85	6616	10	0	MH106727.1	99.43	4560	26	0	A
Spain	EU330453.1	99.83	6616	11	0	EU330452.1	99.52	4543	22	0	A
Sweden	EU330450.1	99.80	6616	13	0	EU330451.1	99.44	4533	25	0	A
Yugoslavia	KX665536.1	99.82	6616	12	0	KX665537.1	98.84	4560	32	1	A
Japan	D84410.1	99.37	6616	42	0	D84411.1	98.47	4560	70	0	A
France (Pithiviers)	HM126464.1	99.41	6616	39	0	HM117903.1	98.42	4560	72	0	A(P)
China	KM434313.1	99.43	6616	38	0	KM434314.1	95.66	4563	195	1	B
France	X05147.1	98.46	6616	102	0	X04197.1	95.35	4563	209	1	B

[a] Indicates NCBI Accession number (#) used as reference; [b] The designation of A- versus B-type is based on sequence differences within the coat protein gene encoded on RNA 2. Additionally, the presence of RNA 5 contributes to the designation of A(P), originally denoting strains from the Pithiviers region of France.

Figure 3. Capped transcripts RNAs 1 and 2 produced from USA BNYVV. (**A**) cDNA clone constructs are infectious in *Chenopodium quinoa*. Transcripts RNAs 1 + 2 combined and rub-inoculated to *C. quinoa* leaves induced chlorotic local lesions from 5 days postinoculation (panels 1–2). (**B**) Total proteins were extracted from local lesions and viral coat protein (CP) was detected by Western blot using anti-BNYVV antiserum (lanes 1–2). Membrane staining (MS) with Coomassie brilliant blue to provide a loading standard. Noninoculated (NI) plants were used as controls. 1 and 2 represent two plants inoculated with RNAs 1 + 2, respectively.

3.4. Novel Virus Discovery in Sugar Beet through RNAseq

The RNA prepared from the enriched virus fraction of Sample S3 possessed an abundant RNA species of ~1.2 kb, an unanticipated outcome of the project. Subsequently, standard total RNA extraction from infected plants and quality analysis by agarose gel electrophoresis revealed the RNA to be visible in agarose gel analysis even without virus particle enrichment (Figure 5). Application of RNAseq to the enriched Sample S3 resulted in greater than 50% of the total reads being assigned to this RNA. Agarose gel electrophoresis of the 1.2 kb RNA suggested that it was a mixture of two isoforms, an observation corroborated by RNAseq reads that were able to be assembled into two related variants (designated BvSatVirus1A and BvSatVirus1B; GenBank accessions MT227166 and MT227167, respectively; Figure S2). A single major ORF present on both RNA isoforms encoded a predicted amino acid sequence with weak similarity to satellite maize white line mosaic virus (MWLMV) [36], a result consistent with the apparent size of the RNA and the recovery of the RNA via virus enrichment.

Additional reads within Sample S3 revealed the presence of the expected sugar beet viruses noted previously, as well as a potential variant of tobacco necrosis virus A (TNV-A; Meulewaeter et al. [37]; Figure 4). Detailed analysis revealed the putative virus sequence (GenBank accession MT227163) to be a potential hybrid between olive mild mosaic virus (OMMV; Cardoso et al. [38]) or olive latent virus-1 (OLV-1; [39]) from the 5′-half through the p8 and p6 genes and TNV-A (CP and 3′-UTR; Figure 4 and Table 3) for a total length of 3682 nt. Uncapped transcript RNA synthesized from a cloned copy of the sequence was demonstrated to be biologically active, producing necrotic local lesions on *C. quinoa*, characteristic of members of the *Alphanecrovirus* group (Figure S3). A more complete characterization of this putative OMMV/TNV-A hybrid and the satellites discovered in this work is in progress in our laboratory (Weiland et al., in preparation). Finally, Sample S4 additionally produced RNAseq reads, enabling the assembly of a variant of satellite tobacco necrosis virus C (98% query coverage possessing 81.4% nt identity with sTNV-C; Accession NC_043430.1).

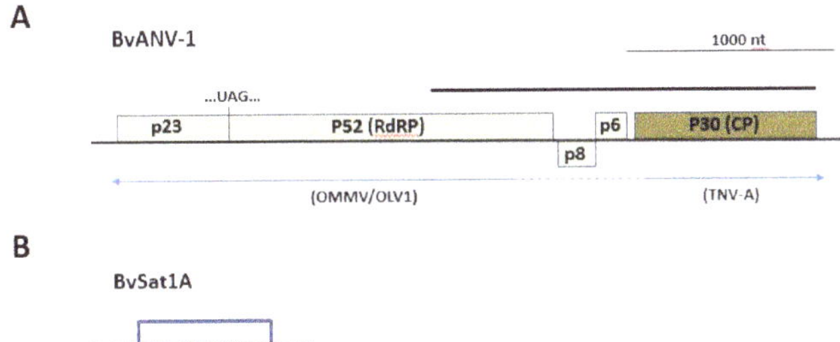

Figure 4. Genome map approximating the organization and size of putative novel viruses associated with sugar beet. (**A**) In the novel *Alphanecrovirus* (BvANV-1), the 5′-proximal ORF is interrupted by an amber stop codon (UAG), as is characteristic for several members of the *Tombusviridae,* and its predicted protein product possesses, along with the predicted movement proteins p8 and p6, greatest similarity to those of the closely related viruses OLV-1 and OMMV. The predicted coat protein (P30-CP) is more closely related to that of TNV-A. Both the 5′- and 3′-UTRs possess the greatest similarity to those from TNV-A. The solid bar above the map represents the single amplicon produced within the study, validating the integrity of the discovered viral genome. (**B**) Approximate genome size and predicted coding information for satellite virus BvSat1A. A single ORF encodes a putative protein with similarity to satellite maize white line mosaic virus.

Table 3. Percent identity between the putative novel Necrovirus genome discovered in sugar beet and domains encoded within the genomes of established Alphanecroviruses.

Virus [a]	Accession # [b]	5′ UTR [c]	P23	P52	P8	P6	CP (P30)	3′ UTR
TNV-A	GCA_000857065.1	93.22	87.13	94.24	89.04	98.21	93.04	78.04%
OMMV	GCF_000858865.1	86.44	96.04	95.01	93.15	98.21	54.13	76.74%
OLV-1	GCF_000855965.1	83.05	90.10	95.59	95.89	100.00	49.07	75.80%
CSNV	MF125267.1	89.83	86.63	93.67	86.30	98.21	51.65	75.46%
PoNV	NC_029900.1	70.00	86.63	89.83	73.97	96.43	50.84	76.05%

[a] Current members of the *Alphanecrovirus* genus by virtue of primary genome and encoded protein sequence structure; [b] Indicates NCBI Accession number (#) used as reference; [c] Similarity (%) of the 5′ and 3′ UTR compared at the nucleotide level. Similarity of ORFs for putative proteins P23, P52, P8, P6, and P30 were compared at the amino acid sequence level.

4. Discussion

Since the advent of the polymerase chain reaction in the mid-1980s, applications of the technology have revolutionized the collection and analysis of DNA and RNA sequence information. RNAseq is one such application that combines high-throughput, short sequence reads of randomly-generated RT-PCR products with contemporary computational power, resulting in "shotgun" sequencing of the RNA population comprising a given sample [40]. Following on published successes using the method to examine viral sequences within bulk cellular nucleic acid sample preparations (reviewed by Schmidt [41]), we sought to use virus enrichment as a means to reduce the sequence complexity of the sample while simultaneously providing a cleaner preparation of RNA from sugar beet roots, which can possess compounds inhibitory to many molecular biology procedures. As expected, known sugar beet viruses garnered the greatest number of reads within the samples, and near-complete genome sequences were obtained. Nevertheless, terminal sequences were, in most cases, absent from the assemblies, a general phenomenon characterizing the technique [42]. Those viruses to which the greatest numbers of reads were assigned were predicted to be present in the soil samples used for virus-baiting based on the soil's cropping history, although the relative number of reads varied between samples submitted and between viruses within a given sample. As the viral RNA was prepared from rhizomania-susceptible sugar beet seedlings harvested in bulk, it was anticipated that this bait would recover a diverse mixture of viruses, and this was born out by our results. By contrast, Sample S6 represents RNA obtained from a single sugar beet plant harvested from a late-season production field and expressing classic rhizomania disease. Viral sequence reads in this sample were dominated by those for BNYVV as compared to bait-plant samples. This is consistent with the fact that Sample S6 was biased, being selected on account of its disease symptoms, as compared to asymptomatic seedlings being bulked from infested soil in the remaining samples.

The results of the present study confirm the power of NGS technologies as applied to RNA-based viral pathogens of sugar beet. Over 98% of the individual genomes of BNYVV, BSBMV, and BBSV were obtained from the read assemblies, with greater than 99% identity in the nucleotide sequence with the closest sequenced relative. A comparison of the assembled sequences for each virus across sample locations within the US indicates a general homogeneity of the sequence. Thus, ~0.13% and 0.2% nt differences between RNA1 and RNA2 sequences were observed, respectively, within the US versus up to 5-fold greater differences between A-type isolates from around the globe (Table 2). An exception to this observation worldwide is nucleotides encoding the "tetrad" of AAs within the p25 protein produced from RNA 3. The nucleotides at this location are considered to be among the most hypervariable within all eukaryotic viruses [25,34,43], a feature observed in our own recent study [17]. Moreover, recent associative [16,17,34] and functional [5,44] evidence suggests that variability in this region may account for the ability for some strains of the virus to circumvent dominant resistance genes in the sugar beet crop. As additional cases exist where a specific tetrad has been observed in both Rz-gene controlled and Rz-gene breaking isolates (e.g., tetrad ACHG [17]), it is possible that other changes in the genome may operate in conjunction with mutations in the tetrad motif or independent of this element in compromising host resistance. The related BSBMV, found only in the US to date, remains a concern as it is not controlled by Rz1 even as it produces only a mild mosaic disease and not the yield losses associated with rhizomania disease [6]. Finally, it was evident that a geographical difference exists for *p25* sequences between isolates in the central states of the US and those existing west of the Rocky Mountain range. This is seen by clustering of the *p25* gene sequence of isolates from Texas, Minnesota, and North Dakota on one branch, separated by additional nodes in the tree from a cluster of isolates obtained from California and Idaho. These specific groupings are also seen for the *p31* sequence, but there are fewer differences between the strains. For both genes, the potential significance of this observation on disease development, vector interaction, or viral fitness remains to be determined. The ability, therefore, to rapidly obtain full genome sequences of BNYVV and other viruses from roots of symptomatic sugar beet will facilitate the detection of other candidate changes conditioning RB in this virus, as well as other viruses or virus variants that may impact the expression of rhizomania disease.

The validity of using RNAseq for the examination of existing, and the discovery of novel, viruses of sugar beet was confirmed through three means in the current study. First, the sequences of BNYVV RNAs 3 and 4 obtained through RNAseq herein were shown to exhibit the closest similarity to archived RNA 3 and RNA 4 sequences from the US that had been obtained in previous studies using standard reverse transcription PCR methods (Figure 2). Several prior investigations utilized sequences of RNAs 3 and 4 for the purposes of categorizing the genetic diversity of BNYVV and ascertaining the origins of the US isolates based on these sequences. We here confirm the grouping proposed by Chiba et al. [25], Schirmer et al. [34], and Zhuo et al. [35], in which US isolates of BNYVV nationwide appear to group with those of the A-types from Italy (Figure 2; Table 2). Although the basis for this apparent relatedness is unknown, it is possible that the virus made its way into US sugar beet production fields through international transit of infested plant material or soil. Transmission of the virus into the US via infected seed can likely be ruled out as no evidence has emerged for the seed-transmission of BNYVV.

Second, the biological validity of the BNYVV sequences obtained was afforded through the construction of clones with demonstrated infectivity based on the consensus sequence from the RNAseq data. The two largest RNAs of the BNYVV genome, RNAs 1 and 2, collectively represent over 70% of the virus genome and encode the replication, packaging, cell–cell movement, and silencing suppression functions of the virus. In contrast to most other biologically-active clones of BNYVV that provide infection through transient genome transcription consequent to *Agrobacterium* infiltration (i.e., "agro-infection"; [44–46]), we chose to employ in vitro production of capped RNA transcriptions as the means to produce inoculum for the infection in recipient cells in a manner more consistent with that found in nature. Along with previous reports of others who used NGS data in the construction of clones from which infectious RNA was produced either through in vitro transcription or via agro-infection [47], our study validates this approach in constructing clones for the study of BNYVV.

Third, the validity of the approach was illustrated by the discovery of a potential novel virus of sugar beet along with a novel satellite virus. The presence of small satellite viruses in sugar beet had not been reported in the USA prior to this study and was unexpected. Analysis of the putative satellite virus genome revealed two variants of a closely-related sequence, both encoding proteins with similarity to themselves and with weak similarity to that encoded by satellite MWLMV, a satellite virus requiring a member of the *Tombusviridae* for its replication [36]. The RNAs differed in size by 157 nt, consistent with the apparent band doublet in a non-denaturing agarose gel (Figure 5). Differences in the size of the two molecules, as predicted from their assembled sequence from that predicted by their electrophoretic migration, may reflect the presence of additional sequences present on those isoforms not captured by RNAseq or conformational aspects of the RNA, resulting in migration anomalies, the possibilities of which are under current investigation.

Further strengthening the validity of this approach in novel virus discovery in sugar beet, the sample in which the satellite virus was present also harbored a previously undocumented variant sequence of TNV-A, possessing a genome organization and gene similarities aligning it with plant Alphanecroviruses within the family *Tombusviridae*. The sequence revealed features suggestive of a recombinant virus with OMMV/OLV-1 and TNV-A as donor parent viruses (Table 3 and Figure 4). The ORF of the putative virus homologous to the replication-associated gene p23 appears to be derived from OMMV, whereas predicted replication protein p52 and movement-associated genes encoding proteins p8 and p6 possess greater similarity to those from OLV-1. At the same time, the predicted CP (p30), the sole CP for members of that sub-genus, appears to be derived from TNV-A in the novel sequence (Table 3 and Figure 4). Since RNAseq data are based on short reads of ~150–200 bp, it has been argued that artifactual contigs might arise during read assembly, potentially providing a false impression of genetic recombination [48]. In the present study, a PCR reaction using primers positioned within the 3′ end of the p52 gene and the 3′ end of the CP gene (Figure 4) yielded a single amplicon, which, when cloned and sequenced, was shown to be homogeneous in sequence and represented the sequence generated by RNAseq. Additionally, within the RNAseq data, no additional "orphaned" ORFs representing other members of the Alphanecroviruses were observed, suggesting that templates of established members of this virus group were absent and could therefore not contribute to the production of artifactual hybrid contigs either through PCR or in silico assembly. Finally, a subsequent genome length amplicon was produced from which RNA was capable of inducing characteristic local lesions on *C. quinoa* (Figure S3). As the sequence of this amplicon matched that obtained through RNAseq within the study, we propose that it represents a new sugar beet-infecting *Alphanecrovirus* within the family *Tombusviridae*, which we propose to be named BvANV-1. Although Liu et al. [49] previously reported TNV from sugar beet in California, no subsequent analysis was conducted to determine the TNV type or its relatedness to other Alphanecroviruses.

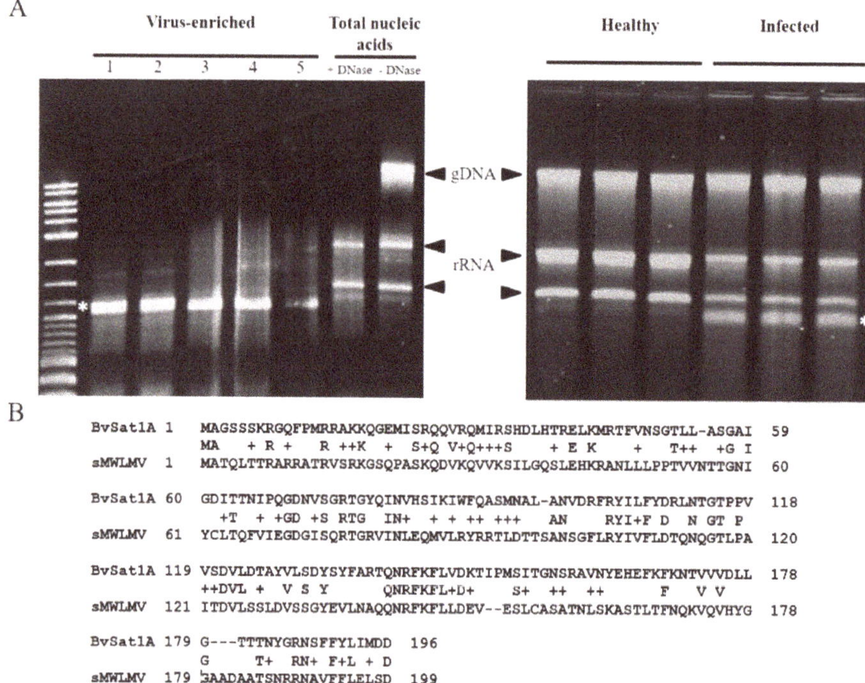

Figure 5. Identification of an abundant satellite virus in sugar beet through virus enrichment and RNAseq. (**A**) Total RNA of healthy sugar beet roots exhibit rRNA and genomic DNA after agarose gel electrophoresis (left panel, -DNase), which can be cleared of DNA (+DNase) prior to RNAseq. Virus enrichment of several infected root samples derived from soils from locations in Minnesota (A, lanes 1–5) results in the persistence of an abundant RNA species of ~1.2 kb present as an apparent doublet band. In the right panel, a high abundance of the satellite virus (*) is evident after agarose gel electrophoresis of total (un-enriched) RNA from healthy and infected plants. (**B**) The single major ORF on the RNA was translated in silico, and the predicted AA sequence (BvSat1A) aligned to that of satellite MWLMV (sMWLMV), the closest relative to the putative novel sugar beet satellite virus in the Genbank database.

A second satellite virus detected in Sample S4 in this study was more closely related to satellite tobacco necrosis virus C, an entity likely associated with helper virus TNV-D or a related Betanecrovirus. Interestingly, Sample S4 harbors BBSV, a well-documented member of that virus subgenus, the US isolate of which was characterized in our laboratory previously [28,32]. Although BBSV is known to have associated satellite RNAs [50], no satellite virus dependent upon BBSV for its replication has, to our knowledge, been reported. The combined compelling results in the present work notwithstanding, for both the novel satellite viruses and candidate helper virus emerging within this study, future investigations will be needed to provide complete sequences and infectious test clones of the genomes. Currently, efforts are underway to test the variant TNV as a helper virus in co-inoculation studies in the presence and absence of the satellite virus.

The NGS approach for detecting known and novel viruses described here and elsewhere provides clear advantages over prior methods of assessment. With reduced bias in the viral genomes targeted, the technique allows for the detection of the presence of unanticipated viruses as compared to more common cloning and sequencing strategies. Moreover, sequence reads for all viruses present are generated simultaneously instead of detecting them within separate, sequential analyses. Standard BLAST alignment can then be employed to detect both known and novel viruses, and more sophisticated applications can detect novel virus agents based on likelihood analysis [51]. Nevertheless, some pitfalls of RNAseq exist due to the short reads produced by the method. Given the existence of virus similarity between members of the same virus family and within family recombinants, one must validate that an assembled sequence represents a true contiguous genome segment through thoughtful primer design and subsequent amplification and cloning of longer sequence segments. As an alternative or complement to RNAseq, one might employ NGS of the like currently offered through PacBio or Oxford Nanopore sequencing platforms that produce long sequence reads from individual molecules as a means to validate sequence contigs generated through RNAseq [52]. Irrespective of the employment of long- or short-read sequencing approaches in initial data acquisition, it appears that 5′-RACE and, to a lesser extent, 3′-RACE will continue to be required in the faithful sequence reproduction of viral RNA genomes and subgenomic transcripts, as 5′- and 3′-terminal structures often are absent or highly underrepresented in sequence reads [42]. Nevertheless, the combination and refinement of NGS technologies have already impacted RNA virology to a great extent in microbial, plant, animal, and human virology, including rapid-response diagnostics and viral genotyping, in recent outbreaks of Ebola virus [53], Zika virus [54], and SARS-CoV2 (COVID-19 infection [55]). As exemplified in the sequencing of multiple genomes of BNYVV and related viruses, and of the discovery of the potentially novel satellite viruses and new *Alphanecrovirus* reported here, the application of the methods promise to revolutionize detection of known and novel viruses of sugar beet, a crop of global importance.

Supplementary Materials: The following are available online at http://www.mdpi.com/1999-4915/12/6/626/s1, Figure S1: General genome organization of dominant viruses detected through RNAseq applied to sugar beet roots. Figure S2. Sequence analysis of the novel satellite virus genome discovered in sugar beet using RNAseq. Figure S3. Infection of *C. quinoa* with novel *Alphanecrovirus* synthetic RNA based on RNAseq genome data and detection of infecting virus by ELISA. Table S1. Location and form of samples use in this study. Table S2. Primers used in this study.

Author Contributions: Conceptualization, J.J.W. and M.D.B.; methodology, J.J.W., R.S.P., A.F., D.E.C., M.D.B.; software, R.S.P., D.E.C.; validation, J.J.W., A.F.; formal analysis, J.J.W., R.S.P., A.F., D.E.C.; investigation, J.J.W., A.F.; resources, G.A.S., M.D.B.; data curation, J.J.W.; writing—original draft preparation, J.J.W.; writing—review and editing, M.D.B.; visualization, J.J.W.; supervision, G.A.S., M.D.B.; project administration, M.D.B.; funding acquisition, M.D.B. All authors have read and agreed to the published version of the manuscript.

Funding: This research received no external funding.

Acknowledgments: The authors thank Jonathan Nebuaer for excellent technical assistance and Rebecca Spanner for bioinformatic assistance.

Conflicts of Interest: The authors declare no conflict of interest.

References

1. Gilardi, G.; Garibaldi, A.; Gullino, M.L. Emerging pathogens as a consequence of globalization and climate change: Leafy vegetables as a case study. *Phytopathol. Mediterr.* **2018**, *57*, 146–152.
2. Jones, R.A.C.; Naidu, R.A. Global dimensions of plant virus diseases: Current status and future perspectives. *Annu. Rev. Virol.* **2019**, *6*, 387–409. [CrossRef] [PubMed]
3. Friesen, T.L.; Stukenbrock, E.H.; Liu, Z.; Meinhardt, S.; Ling, H.; Faris, J.D.; Rasmussen, J.B.; Solomon, P.S.; McDonald, B.A.; Oliver, R.P. Emergence of a new disease as a result of interspecific virulence gene transfer. *Nat. Genet.* **2006**, *38*, 953–956. [CrossRef] [PubMed]
4. Kado, C.I. Horizontal gene transfer: Sustaining pathogenicity and optimizing host–pathogen interactions. *Mol. Plant Pathol.* **2009**, *10*, 143–150. [CrossRef] [PubMed]
5. Koenig, R.; Loss, S.; Specht, J.; Varrelmann, M.; Lüddecke, P.; Deml, G. A single U/C nucleotide substitution changing alanine to valine in the Beet necrotic yellow vein virus P25 protein promotes increased virus accumulation in roots of mechanically inoculated, partially resistant sugar beet seedlings. *J. Gen. Virol.* **2009**, *90*, 759–763. [CrossRef] [PubMed]
6. Rush, C.M.; Liu, H.Y.; Lewellen, R.T.; Acosta-Leal, R. The continuing saga of rhizomania of sugar beets in the United States. *Plant Dis.* **2006**, *90*, 4–15. [CrossRef] [PubMed]
7. Biancardi, E.; Lewellen, R.T.; De Biaggi, M.; Erichsen, A.W.; Stevanato, P. The origin of rhizomania resistance in sugar beet. *Euphytica* **2002**, *127*, 383–397. [CrossRef]

8.   Canova, A.; Giunchedi, L.; Biancardi, E. History and current status. In *Rhizomania*; Biancardi, E., Tamada, T., Eds.; Springer International Publishing: Basel, Switzerland, 2016; pp. 29–51.

9.   Richards, K.; Tamada, T. Mapping functions on the multipartite genome of Beet necrotic yellow vein virus. *Annu. Rev. Phytopathol.* **1992**, *30*, 291–313. [CrossRef]

10.   Tamada, T. General features of Beet necrotic yellow vein virus. In *Rhizomania*; Biancardi, E., Tamada, T., Eds.; Springer International Publishing: Basel, Switzerland, 2016; pp. 55–83.

11.   Peltier, C.; Hleibieh, K.; Thiel, H.; Klein, E.; Bragard, C.; Gilmer, D. oMlecular biology of the Beet necrotic yellow vein virus. *Plant Viruses* **2008**, *2*, 14–24.

12.   Laufer, M.; Mohammad, H.; Maiss, E.; Richert-Pöggeler, K.; Dall'Ara, M.; Ratti, C.; Gilmer, D.; Liebe, S.; Varrelmann, M. Biological properties of Beet soil-borne mosaic virus and Beet necrotic yellow vein virus cDNA clones produced by isothermal in vitro recombination: Insights for reassortant appearance. *Virology* **2018**, *518*, 25–33. [CrossRef]

13.   Gilmer, D. Molecular biology and replication of Beet Necrotic Yellow Vein Virus. In *Rhizomania*; Biancardi, E., Tamada, T., Eds.; Springer International Publishing: Basel, Switzerland, 2016; pp. 85–107.

14.   Acosta-Leal, R.; Bryan, B.K.; Smith, J.T.; Rush, C.M. Breakdown of host resistance by independent evolutionary lineages of Beet necrotic yellow vein virus involves a parallel C/U mutation in its p25 gene. *Phytopathology* **2010**, *100*, 127–133. [CrossRef] [PubMed]

15.   Bornemann, K.; Hanse, B.; Varrelmann, M.; Stevens, M. Occurrence of resistance-breaking strains of Beet necrotic yellow vein virus in sugar beet in northwestern Europe and identification of a new variant of the viral pathogenicity factor P25. *Plant Pathol.* **2015**, *64*, 25–34. [CrossRef]

16.   Liu, H.Y.; Lewellen, R.T. Distribution and molecular characterization of resistance-breaking isolates of Beet necrotic yellow vein virus in the United States. *Plant Dis.* **2007**, *91*, 847–851. [CrossRef] [PubMed]

17.   Weiland, J.J.; Bornemann, K.; Neubauer, J.D.; Khan, M.F.R.; Bolton, M.D. Prevalence and Distribution of Beet Necrotic Yellow Vein Virus Strains in North Dakota and Minnesota. *Plant. Dis.* **2019**, *103*, 2083–2089. [CrossRef] [PubMed]

18.   Lecuit, M.; Eloit, M. The potential of whole genome NGS for infectious disease diagnosis. *Expert Rev. Mol. Diagn.* **2015**, *15*, 1517–1519. [CrossRef] [PubMed]

19.   Villamor, D.E.V.; Ho, T.; Al Rwahnih, M.; Martin, R.R.; Tzanetakis, I.E. High Throughput Sequencing for Plant Virus Detection and Discovery. *Phytopathology* **2019**, *109*, 716–725. [CrossRef] [PubMed]

20.   Filloux, D.; Dallot, S.; Delaunay, A.; Galzi, S.; Jacquot, E.; Roumagnac, P. Metagenomics Approaches Based on Virion-Associated Nucleic Acids (VANA): An Innovative Tool for Assessing Without A Priori Viral Diversity of Plants. In *Plant Pathology: Techniques and Protocols*; Lacomme, C., Ed.; Springer: New York, NY, USA, 2015; pp. 249–257.

21.   Ma, Y.; Marais, A.; Lefebvre, M.; Theil, S.; Svanella-Dumas, L.; Faure, C.; Candresse, T. Phytovirome Analysis of Wild Plant Populations: Comparison of Double-Stranded RNA and Virion-Associated Nucleic Acid Metagenomic Approaches. *J. Virol.* **2019**, *94*. [CrossRef] [PubMed]

22.   Greninger, A.L. A decade of RNA virus metagenomics is (not) enough. *Virus Res.* **2018**, *244*, 218–229. [CrossRef] [PubMed]

23.   Lane, L.C. Propagation and purification of RNA plant viruses. In *Methods in Enzymology*; Weissbach, A., Weissbach, H., Eds.; Elsevier: Amsterdam, The Netherlands, 2004; Volume 118, pp. 687–696.

24.   Weiland, J.J.; Edwards, M.C. Evidence That the αa Gene of Barley Stripe Mosaic Virus Encodes Determinants of Pathogenicity to Oat (Avena sativa). *Virology* **1994**, *201*, 116–126. [CrossRef] [PubMed]

25.   Chiba, S.; Kondo, H.; Miyanishi, M.; Andika, I.B.; Han, C.; Tamada, T. The evolutionary history of Beet necrotic yellow vein virus deduced from genetic variation, geographical origin and spread, and the breaking of host resistance. *Mol. Plant Microbe Interact.* **2011**, *24*, 207–218. [CrossRef] [PubMed]

26.   Petty, I.T.D.; Donald, R.G.K.; Jackson, A.O. Multiple Genetic Determinants of Barley Stripe Mosaic Virus Influence Lesion Phenotype on Chenopodium amaranticolor. *Virology* **1994**, *198*, 218–226. [CrossRef] [PubMed]

27.   Edwards, M.C.; Weiland, J.J.; Todd, J.; Stewart, L.R. Infectious Maize rayado fino virus from Cloned cDNA. *Phytopathology* **2015**, *105*, 833–839. [CrossRef] [PubMed]

28.   Weiland, J.J.; Van Winkle, D.; Edwards, M.C.; Larson, R.L.; Shelver, W.L.; Freeman, T.P.; Liu, H.Y. Characterization of a U.S. isolate of *Beet black scorch virus*. *Phytopathology* **2007**, *97*, 1245–1254. [CrossRef] [PubMed]

29. Heidel, G.B.; Rush, C.M.; Kendall, T.L.; Lommel, S.A.; French, R.C. Characteristics of Beet Soilborne Mosaic Virus, a Furo-like Virus Infecting Sugar Beet. *Plant Dis.* **1997**, *81*, 1070–1076. [CrossRef] [PubMed]

30. Henry, C.M.; Jones, R.A.C.; Coutts, R.H.A. Occurrence of a soil-borne virus of sugar beet in England. *Plant Pathol.* **1986**, *35*, 585–591. [CrossRef]

31. Koenig, R.; Pleij, C.W.; Beier, C.; Commandeur, U. Genome properties of beet virus Q, a new furo-like virus from sugarbeet, determined from unpurified virus. *J. Gen. Virol.* **1998**, *79*, 2027–2036. [CrossRef]

32. Weiland, J.J.; Larson, R.L.; Freeman, T.P.; Edwards, M.C. First Report of Beet black scorch virus in the United States. *Plant Dis.* **2006**, *90*, 828. [CrossRef]

33. Lee, L.; Telford, E.B.; Batten, J.S.; Scholthof, K.B.G.; Rush, C.M. Complete nucleotide sequence and genome organization of *Beet soilborne mosaic virus*, a proposed member of the genus Benyvirus. *Arch. Virol.* **2001**, *146*, 2443–2453. [CrossRef] [PubMed]

34. Schirmer, A.; Link, D.; Cognat, V.; Moury, B.; Beuve, M.; Meunier, A.; Bragard, C.; Gilmer, D.; Lemaire, O. Phylogenetic analysis of isolates of *Beet necrotic yellow vein virus* collected worldwide. *J. Gen. Virol* **2005**, *86 Pt 10*, 2897–2911. [CrossRef]

35. Zhuo, N.; Jiang, N.; Zhang, C.; Zhang, Z.Y.; Zhang, G.Z.; Han, C.G.; Wang, Y. Genetic diversity and population structure of *Beet necrotic yellow vein virus* in China. *Virus Res.* **2015**, *205*, 54–62. [CrossRef]

36. Zhang, L.; Zitter, T.A.; Palukaitis, P. Helper virus-dependent replication, nucleotide sequence and genome organization of the satellite virus of maize white line mosaic virus. *Virology* **1991**, *180*, 467–473. [CrossRef]

37. Meulewaeter, F.; Seurinck, J.; Emmelo, J.V. Genome structure of tobacco necrosis virus strain A. *Virology* **1990**, *177*, 699–709. [CrossRef]

38. Cardoso, J.M.S.; Félix, M.R.; Clara, M.I.E.; Oliveira, S.J. The complete genome sequence of a new necrovirus isolated from Olea europaea L. *Arch. Virol.* **2005**, *150*, 815–823. [CrossRef] [PubMed]

39. Félix, M.R.; Cardoso, J.; Varanda, C.M.; Oliveira, S.; Clara, M.I.E. Complete nucleotide sequence of an Olive latent virus 1 isolate from olive trees. *Arch. Virol.* **2005**, *150*, 2403–2406. [CrossRef] [PubMed]

40. Wang, Z.; Gerstein, M.; Snyder, M. RNA-Seq: A revolutionary tool for transcriptomics. *Nat. Rev. Genet.* **2009**, *10*, 57–63. [CrossRef] [PubMed]

41. Schmidt, C. The virome hunters. *Nature* **2018**, *36*, 916–919. [CrossRef] [PubMed]

42. Corney, D.C.; Basturea, G.N. RNA-Seq using next generation sequencing. *Mater. Methods* **2013**, *3*. [CrossRef]

43. Acosta-Leal, R.; Fawley, M.W.; Rush, C.M. Changes in the intraisolate genetic structure of *Beet necrotic yellow vein virus* populations associated with plant resistance breakdown. *Virology* **2008**, *376*, 60–68. [CrossRef]

44. Liebe, S.; Wibberg, D.; Maiss, E.; Varrelmann, M. Application of a Reverse Genetic System for Beet Necrotic Yellow Vein Virus to Study Rz1 Resistance Response in Sugar Beet. *Front. Plant Sci.* **2020**, *10*, 1703. [CrossRef] [PubMed]

45. Delbianco, A.; Lanzoni, C.; Klein, E.; Rubies Autonell, C.; Gilmer, D.; Ratti, C. Agroinoculation of Beet necrotic yellow vein virus cDNA clones results in plant systemic infection and efficient Polymyxa betae transmission. *Mol. Plant Pathol.* **2013**, *14*, 422–428. [CrossRef]

46. Jiang, N.; Zhang, C.; Liu, J.Y.; Guo, Z.H.; Zhang, Z.Y.; Han, C.G.; Wang, Y. Development of Beet necrotic yellow vein virus-based vectors for multiple-gene expression and guide RNA delivery in plant genome editing. *Plant Biotechnol. J.* **2019**, *17*, 1302–1315. [CrossRef] [PubMed]

47. Pasin, F.; Menzel, W.; Daròs, J.A. Harnessed viruses in the age of metagenomics and synthetic biology: An update on infectious clone assembly and biotechnologies of plant viruses. *Plant Biotechnol. J.* **2019**, *17*, 1010–1026. [CrossRef] [PubMed]

48. Yang, Y.; Smith, S.A. Optimizing de novo assembly of short-read RNA-seq data for phylogenomics. *BMC Genom.* **2013**, *14*, 328. [CrossRef] [PubMed]

49. Liu, H.Y.; Duffus, J.E.; Wisler, G.C. *Etiology of Vascular Necrosis Syndrome of Sugarbeet*; Rothamsted Research: Harpenden, UK, 1996; pp. 161–164.

50. Guo, L.H.; Cao, Y.-H.; Li, D.W.; Niu, S.N.; Cai, Z.N.; Han, C.G.; Zhai, Y.F.; Yu, J.-L. Analysis of Nucleotide Sequences and Multimeric Forms of a Novel Satellite RNA Associated with Beet Black Scorch Virus. *J. Virol* **2005**, *79*, 3664–3674. [CrossRef] [PubMed]

51. Ho, T.; Tzanetakis, I.E. Development of a virus detection and discovery pipeline using next generation sequencing. *Virology* **2014**, *471–473*, 54–60. [CrossRef] [PubMed]

52. Weirather, J.L.; De Cesare, M.; Wang, Y.; Piazza, P.; Sebastiano, V.; Wang, X.J.; Buck, D.; Au, K.F. Comprehensive comparison of Pacific Biosciences and Oxford Nanopore Technologies and their applications to transcriptome analysis. *F1000Res* **2017**, *6*, 100. [CrossRef] [PubMed]
53. Matranga, C.B.; Andersen, K.G.; Winnicki, S.; Busby, M.; Gladden, A.D.; Tewhey, R.; Stremlau, M.; Berlin, A.; Gire, S.K.; England, E.; et al. Enhanced methods for unbiased deep sequencing of Lassa and Ebola RNA viruses from clinical and biological samples. *Genome Biol.* **2014**, *15*, 519. [CrossRef] [PubMed]
54. Metsky, H.C.; Matranga, C.B.; Wohl, S.; Schaffner, S.F.; Freije, C.A.; Winnicki, S.M.; West, K.; Qu, J.; Baniecki, M.L.; Gladden-Young, A.; et al. Zika virus evolution and spread in the Americas. *Nature* **2017**, *546*, 411–415. [CrossRef] [PubMed]
55. Wu, F.; Zhao, S.; Yu, B.; Chen, Y.M.; Wang, W.; Song, Z.G.; Hu, Y.; Tao, Z.W.; Tian, J.H.; Pei, Y.Y.; et al. A new coronavirus associated with human respiratory disease in China. *Nature* **2020**, *579*, 265–269. [CrossRef] [PubMed]

Article

Family Level Phylogenies Reveal Relationships of Plant Viruses within the Order Bunyavirales

Venura Herath [1,2], Gustavo Romay [1], Cesar D. Urrutia [1] and Jeanmarie Verchot [1,*]

[1] Department of Plant Pathology & Microbiology, Texas A&M University, College Station, TX 77802, USA; venura.herath@tamu.edu (V.H.); gustavo.romay@tamu.edu (G.R.); curru001@tamu.edu (C.D.U.)

[2] Department of Agriculture Biology, Faculty of Agriculture, University of Peradeniya, Peradeniya 20400, Sri Lanka

* Correspondence: jm.verchot@tamu.edu; Tel.: +1-979-845-1788

Received: 19 August 2020; Accepted: 8 September 2020; Published: 10 September 2020

Abstract: *Bunyavirales* are negative-sense segmented RNA viruses infecting arthropods, protozoans, plants, and animals. This study examines the phylogenetic relationships of plant viruses within this order, many of which are recently classified species. Comprehensive phylogenetic analyses of the viral RNA dependent RNA polymerase (RdRp), precursor glycoprotein (preGP), the nucleocapsid (N) proteins point toward common progenitor viruses. The RdRp of *Fimoviridae* and *Tospoviridae* show a close evolutional relationship while the preGP of *Fimoviridae* and *Phenuiviridae* show a closed relationship. The N proteins of *Fimoviridae* were closer to the *Phasmaviridae*, the *Tospoviridae* were close to some *Phenuiviridae* members and the *Peribunyaviridae*. The plant viral movement proteins of species within the *Tospoviridae* and *Phenuiviridae* were more closely related to each other than to members of the *Fimoviridae*. Interestingly, distal ends of 3′ and 5′ untranslated regions of species within the *Fimoviridae* shared similarity to arthropod and vertebrate infecting members of the *Cruliviridae* and *Peribunyaviridae* compared to other plant virus families. Co-phylogeny analysis of the plant infecting viruses indicates that duplication and host switching were more common than co-divergence with a host species.

Keywords: *Bunyavirale*; RNA virus; emerging virus; virus evolution; plant virus; cophylogeny; hallmark genes

---

1. Introduction

Viruses in the order *Bunyavirales* infect arthropods, plants, protozoans, and vertebrates. Their RNA genomes are segmented and exhibit negative or ambisense polarity. Each virus species has a fixed number of genome segments which range from two to eight, with plant viruses having the largest numbers of segments. The nucleotide sequences at the 3′ and 5′ terminus of each genome segment are complementary and form panhandle structures for stability. Their RNA segments are mostly coated in nucleocapsid proteins and further encapsulated in an envelope derived from its host cell.

Bunyavirales is a recently established taxonomic order that encompasses twelve families comprising 46 genera [1,2]. Four families contain members that cause life-threatening diseases in humans: *Hantaviridae*, *Nairoviridae*, *Peribunyaviridae* and *Phenuiviridae* [1,2]. These families include the species *Bunyamwera virus (BUNV)*, *Crimean-Congo haemorrhagic fever virus (CCHFV)*, *Hantaan virus (HTNV)*, *La Crosse virus (LACV)*, *Rift Valley fever virus (RVFV)*, *Severe fever with thrombocytopenia syndrome virus (SFTSV)*, and *Sin Nombre virus (SNV)*. Three families within *Bunyavirales* contain members that infect plants as their primary host: *Fimoviridae*, *Phenuiviridae*, and *Tospoviridae*. Within these families, there is one genus of plant infecting viruses: *Emaravirus*, *Tenuivirus*, and *Orthotospovirus*, respectively. Across *Bunyavirales*, viruses can have three segments of negative-sense or ambisense RNA that are named according to their relative length. These segments are known as large (L), medium (M), and small (S) which encode the viral RNA dependent RNA polymerase (RdRp), a polyprotein precursor glycoprotein (preGP) that is co-translationally cleaved into two mature glycoproteins (Gn and Gc), and the nucleocapsid (N) protein, respectively. Among the plant viruses with more than three genome segments where "x" equals the total number of segments, each segment is numbered from RNA1 to RNAx and are assigned sequentially to each segment in order of decreasing nucleotide length. A novel genus that is tentatively named *Coguvirus* has a bi-partite genome, lacks an external envelope and, is proposed to the order *Bunyavirales* [3,4].

The origins and evolutionary history of plant viruses within *Bunyavirales* are unclear. While phylogenetic studies suggest common ancestral origins of vertebrate and arthropod infecting viruses, the ancestral lineages of plant viruses within this taxonomic order have not been well studied [5,6]. Extensive sampling of arthropods (crustaceans, centipedes, insects, and spiders) have revealed new species of negative-sense RNA viruses, and many appear to be ancestral to viruses that cause diseases in vertebrate hosts [5,7–10]. Koonin and Dolja (2014) coined the term "hallmark genes" referring to viral genes that encode the necessary apparatus of viral replication and encapsidation and provide important clues about the evolutionary origins of disease-causing viruses. Studies of hallmark genes provide insight into the shared and conserved domain modules that are used in classification schemes to understand common evolutionary histories [11]. While there are extensive reports on the evolutionary relationships among positive-strand RNA and double-strand RNA viruses built on the analyses of viral hallmark genes, less is known about the evolutionary connections among the hallmark genes of plant-infecting viruses with negative sense or ambisense genomes, especially within *Bunyavirales* [4,12–14].

Recent research in the field of virus metagenomics has expanded the list of new plant-infecting species within *Bunyavirales*, which has contributed to the recent reorganization of families within this taxonomic order [2]. This study examines the phylogenetic lineages and host associations of recently discovered plant-infecting viruses within *Bunyavirales* by examining the shared and conserved hallmark genes among arthropod, plant, protozoan, and animal-infecting counterparts. This study also includes analysis of plant viral movement proteins which represent important changes in virus evolution from deeply rooted ancestral viruses.

2. Materials and Methods

2.1. Phylogenetic Analysis of Bunyavirales

We retrieved RdRp, preGP, N, and MP sequences from the NCBI protein archive (Supplementary Table S1). We used the updated taxonomy of the order *Bunyavirales* by the International Committee on Taxonomy of Viruses (ICTV) [2] as a guide to retrieve sequences of each representative virus species. Retrieved sequences were aligned using MAFFT ver. 7 [15–17] using E-INS-i algorithm. Ambiguously aligned regions were removed using the trimming mode ML_Automated1 of TrimAl ver. 1.3 wrapper embedded in TBTools ver. 1.0 [18,19]. ProtTest ver. 3.4.2 was used to determine the best candidate of the amino acid substitution models for all sequence alignments. LG+I+G+F, LG+G+F, LG+G and LG+I+G+F amino acid replacement models were used for the phylogenetic analysis of RdRp, NC, GP and MP respectively [20]. Phylogenetic trees were generated using PhyML program ver. 3.1 with the maximum likelihood (ML) approach embedded in SeaView ver. 5.0.4 [21,22]. Tree searching was employed using the nearest neighbor interchange (NNI) search strategy. Branch support was computed using an approximate likelihood ratio test (aLRT) with the Shimodaira–Hasegawa-like (SH) procedure. Phylogenetic trees were visualized using iTOL server ver 5.6 [23,24]. Images were compiled using Adobe Photoshop CC (ver. 21.2.0).

2.2. Analysis of the Untranslated Regions (UTRs) of RNA Segments

The 3′ and 5′ UTR regions of viral segments were extracted using NCBI nucleotide database (Supplementary Tables S2 and S3). Sequences were manually checked using RNAfold ver. 2.4.14 [25] plugin built into Geneious Prime ver. 2020.2 for sequence quality and completeness. Then the first 20 nucleotides were extracted using the same program. Sequence logos were created using the WebLogo 3 server [26,27]. Images were compiled using Adobe Photoshop CC (ver. 21.2.0).

2.3. Pairwise Sequence Alignment and Identity Score Calculation

For calculating identity scores of MP amino acid sequences, pairwise sequence alignment was performed using the software Sequence Demarcation Tool (SDT) v. 1.2 [28].

2.4. Co-Phylogenetic Analysis

Cophylogenetic relationships between families and their natural hosts were analyzed with event-based co-phylogeny analysis tool Jane ver. 4.01 [29]. Phylogenetic relationships among the hosts were obtained from the NCBI Taxonomy browser [30]. The host information was obtained from the Virus–Host DB [31] and available literature [32] (Supplemental Table S4). Phylogenies of virus families were conducted based on the RdRp protein sequences as described above. Viruses without host information were excluded from the analysis. The following cost scheme was used for the analysis in Jane; co-divergence = 0, duplication = 1, host switch = 1, loss = 1, failure to diverge = 1. The number of generations and the population size was both set to 100. In order to visualize the taxonomic relationships between plant and insect taxa, we used concatenated genomic segments (L, M, S, and RNA 4 segments) containing four hallmark genes (RdRp, NC, GP, and MP) of plant viruses. Viruses with missing segments and incomplete sequences were excluded from the analysis. The sequence concatenation was carried out using Geneious Prime version 2020.2.1. Concatenated sequences were aligned using MAFFT server version 7 [17] using E-INS-i method [16]. A neighborhood joining tree was generated using the conserved sites (1800 nts) using Jukes–Cantor substitution model with 1000 bootstraps using MAFFT server version 7 [16]. Plant host taxonomies were obtained from APWeb version 14 [33,34]. The resulting phylogenetic tree was visualized and color-coded in iTOL server version 5.6 [23,24]. Image compilation was carried out in Photoshop CC version 21.2.0 and Illustrator version 24.2.3.

3. Results

3.1. Phylogeny and Domain Analysis of RNA-Dependent RNA Polymerase (RdRp), Glycoprotein Precursor (preGP), Nucleocapsid Proteins (N), and Movement Proteins (MP) of Bunyavirales Members

3.1.1. Phylogeny of RdRp

For all negative-strand RNA viruses in the order *Bunyavirales*, RNA1, (or the L segment) is the longest and encodes RdRp. The RdRp sequences for 253 species belonging to arthropod, plant, protozoan, and vertebrate infecting viruses within *Bunyavirales* were compiled (Supplementary Table S1) to build an ML phylogeny. The ML tree in Figure 1 covers 12 families and one unassigned species and, has three deeply rooted clades with viruses of insect hosts at the basal position as reported in Guterres et al. (2017) [6]. Within these three clades are six major lineages that we identified as groups I through VI (Figure 1). These groups are recognized based on the cluster of branches emanating from the most distant node, suggesting a common lineage progenitor. These lineage groups are supported by their primary hosts (protozoa, plant, arthropod, and vertebrate). Except for group II, all other groups contain families that infect vertebrates and/or invertebrates. Notably, the species *Chilibre phlebovirus* (CHIV) is classified by the ICTV as a member of the family *Phenuiviridae* but the ML tree indicates that the RdRp is in the lineage group I with *Peribunyaviridae* family and clusters with the *Pacuvirus* and *Herbevirus* genera. This unusual relationship, verified using the aLRT-SH test (Supplementary Figure S1), suggests that the taxonomic assignment of CHIV may be erroneous.

Viruses of lineage groups I, II, and III traces to one deeply rooted clade (Figure 1) identified by Guterres et al. (2017) as a Bunyavirus-like supergroup [6]. The deepest root of this clade leads to group III viruses that include the *Orthophasmavirus*, *Jonvirus*, and *Feravirus* genera. The *Orthohantavirus* genus is the next bifurcation in Group III. Within this large clade is another deep root that bifurcates to group II plant-infecting *Orthotospovirus* and *Emaravirus* and the larger group I *Orthobunyavirus* and *Lincruvirus* genera. The species *Crustacean lincruvirus* is at the root of the group I *Orthobunyavirus* lineage [9]. Emaraviruses are vectored by mites and orthotospoviruses are vectored by thrips [12,35,36].

The next deeply rooted clade includes groups IV and V, *Arenavirus* and *Nairovirus*. This is known as the arenanairo-like virus superclade according to Guterres et al. (2017) [6]. The invertebrate-infecting species *Myriapod hubavirus*, *Haartman hartmanvirus*, and *Striated antennavirus* are at the deepest root supporting lineage group IV viruses. The invertebrate-infecting species *Millipede wumivirus* is at the deepest root supporting lineage group V.

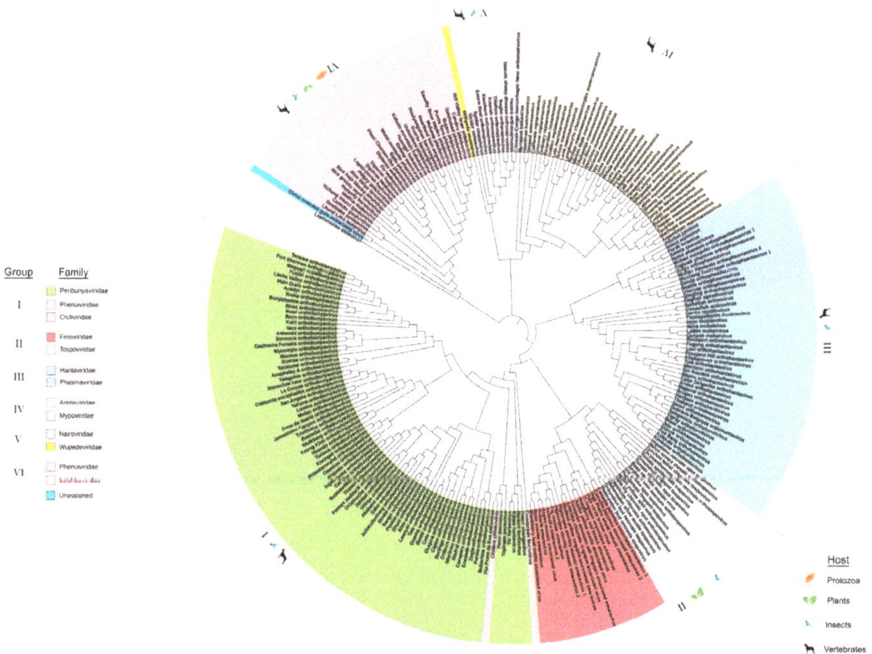

Figure 1. Maximum likelihood phylogenetic tree of the amino acid sequences of the RNA-dependent RNA polymerase (RdRp). The virus families are color-coded and the hosts for viruses within each group are indicated in the outermost circle. The six groups are identified in the legend and the boundaries of these groups are indicated in the outer ring of the phylogeny. Group I: *Peribunyaviridae*, *Phenuiviridae*, and *Cruliviridae*. Group II: *Fimoviridae* and *Tospoviridae*. Group III: *Hantaviridae* and *Phasmaviridae*. Group IV: *Arenaviridae* and *Mypoviridae*. Group V: *Nairoviridae* and *Wupedeviridae*. Group VI: *Phenuiviridae*, *Leishbuviridae*, and unassigned species. Clade validation is based on the approximate likelihood ratio test (aLRT)-Shimodaira–Hasegawa (SH)-like test values.

The third major branch has the invertebrate-infecting species *Leptomonas shilevirus* and *Laurel Lake virus* at the deepest node. The plant-infecting genera *Tenuivirus* and *Coguvirus* and the insect-infecting genus *Goukuvirus* are the closest relatives to these invertebrate-infecting genera [3]. While Guterres et al. (2017) identified this as a phlebo-like virus superclade, the *Phlebovirus* genus represents a smaller fraction of viruses within this lineage group with the majority of viruses representing plant, insect, and protist-infecting viruses [6]. The RdRps of the plant-infecting virus species within this phylogeny, like the vertebrate-infecting viruses, appear to have arisen from common progenitor viruses [37].

Considering the close relationship between the families *Fimoviridae* and *Tospoviridae*, we carefully examined an alignment of their RdRp sequences. Crystal structures of several members of the order *Bunyavirales* have been used to identify functional motifs and similarities across species within the order, and these reports informed this analysis [38–40]. The N-terminal domain harbors the endonuclease activity required for cap-snatching processes (Figure 2A). The polymerase domain near the C-terminus has motifs preA and A through E which are conserved in linear arrangement and distance (Figure 2B). Members of the *Fimoviridae* and *Tospoviridae* share the conserved motifs (H … PD … D/E … K … T/Y … Y) in the endonuclease active center occurring in all families of *Bunyavirales* [35,38,40], but with a few modifications. First, species within the *Orthobunyavirus*, *Orthotospovirus*, *Hantavirus*, and *Phlebovirus* genera have the conserved D/E motif between the H and PD (Figure 2A, position 145 in the alignment) that also occurs in members of the genus *Orthotospovirus* but is missing in members of the genus *Emaravirus* [35,38,40]. The T/Y at position 225 is reported as T/K for members of the *Orthobunyavirus*, *Hantavirus*, and *Phlebovirus*. This alignment shows the T/K is conserved at position 225 for *Fimoviridae* and *Tospoviridae* members. Orthotospoviruses have two added sequences between positions 165 and190, and between 242 and 254 (Figure 2). The C-terminal polymerase domain is highly conserved between *Fimoviridae* and *Tospoviridae*. The motifs preA, A through E have a high proportion of identical and highly conserved residue with only a few minor changes that differentiate members of the genera *Emaravirus* and *Orthotospovirus*. One minor difference occurs in the preA motif at position 1388 to 1390; *Fimoviridae* has a tripeptide that is NxQ while *Tospoviridae* has SMK. In motif A, at position 1452 to 1455, *Fimoviridae* has LSSD and *Tospoviridae has* LSAD. At position 1500 to 1510, which is between motifs A and B, emaraviruses have IxLTDxxN/DxF and orthotospoviruses have VCIPTDIFLNL. Then, at position 1581 in motif C, emaraviruses have S/F/Y while orthotospoviruses have W.

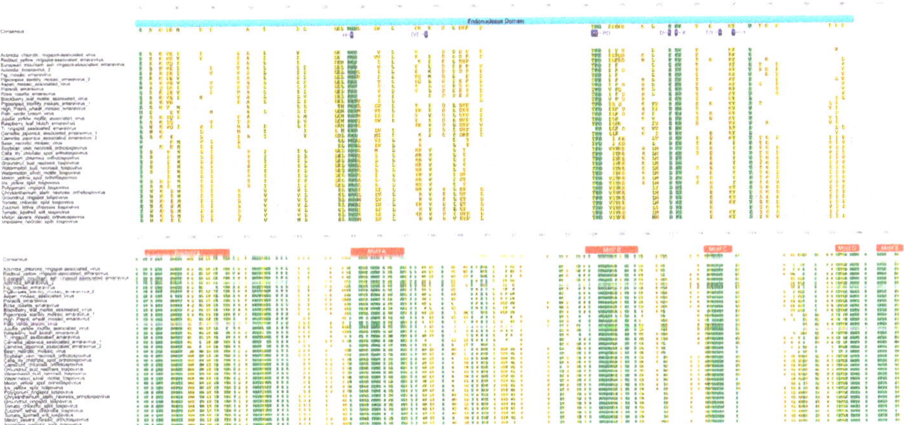

Figure 2. Amino acid alignment showing conserved motifs of the RdRp within *Fimoviridae* and Tospoviridae. (**A**). The endonuclease domain is indicated by pink bar and active site motifs are identified in blue. (**B**). The polymerase function motifs are named in the red bars as preA motif through E motif. The alignment colored based on the sequence similarity.

3.1.2. Phylogeny of preGP

The *Bunyavirales* RNA2 (or M segment) encodes the preGP which is inserted into the host endoplasmic reticulum (ER) and cleaved by the cellular signalase into Gn and Gc [41]. The mature Gn and Gc are required for virus particle budding and entry into target cells. Most virus members also encode one or more nonstructural proteins including the major nonstructural protein (NSm) which are positioned in one of five locations within the RNA2 or M segment (Supplementary Figure S2). In general, the NSm of vertebrate-infecting viruses is required for virus growth in cell cultures but is dispensable for virus replication. For plant-infecting viruses, NSm is required for virus cell-to-cell spread. For *Nairoviridae* and *Peribunyaviridae*, the NSm is part of the polyprotein located between the Gn and Gc regions. For *Phasmaviridae*, NSm is located at the N-terminus of the Gn sequence. For *Tospoviridae*, the NSm is ambisense oriented and located next to the Gc domain. For *Phenuiviridae*, the NSm is an open reading frame nested within the Gn region. Members of the genera *Tenuivirus* and *Emaravirus* lack the NSm coding sequence in RNA2 and M segment (Supplementary Figure S1). The tenuiviruses and emaraviruses have more than three genome segments, and their MP is located on another genome segment.

The preGP phylogeny shows three deeply rooted branches and displays six major lineage groups comprising 11 taxonomic families (Figure 3). These lineage groups are supported by their primary host (vertebrate plant, and insect). The *Leishbunyaviridae*, and unassigned *Coguvirus* were not included because the full-length sequences for the M segment (RNA 2) are not available. The *Gouleako goukuvirus*, which is the type member of the genus *Goukuvirus* (family *Phenuiviridae*), is a deeply rooted branch that precedes the major subclades in groups I, II, and III. Looking at the M segment (or RNA2 segment) for each genus within these virus families, the length of the preGP varies significantly. In group I, *Orthonairovirus* fall into two classes that either contain or lack the NSm within the polyprotein (Supplementary Figure S2). The members of the plant-infecting virus genus *Orthotospovirus* encode NSm in an ambisense direction which does not overlap the glycoprotein precursor. In group II, only members of the genus *Feravirus* contain an NSm sequence, however, this does not overlap the glycoprotein precursor. The plant-infecting members of *Emaravirus* and *Tenuivirus* do not encode NSm (Supplementary Figure S2). Among group III, the NSm adjacent to the Gn domain of the polyprotein for *Orthobunyavirus*, *Shangavirus*, and *Jonvirus*. It is reasonable to suggest that the NSm likely influenced the diversification of some viral preGPs within the ML tree, but given the diversity of the preGPs, there are likely to be other factors affecting their evolution (Supplementary Figure S2).

Two deeply rooted branches lead to the group VI and group V, the primarily arthropod-borne species of *Peribunyaviridae* (*Orthobunyavirus* and *Pacuvirus*) and *Phenuiviridae*. From the *Pacuvirus* branch, there are three major subclades: three species of *Orthobunyavirus* cluster in group VI, the group V *Phenuiviridae* cluster, and the group IV cluster of primarily *Hantaviridae* with sole representatives of *Arenaviridae* and *Mypoviridae* (Figure 3). Among group V, some phleboviruses such as *Rift valley fever virus*, contain the NSm as a nested gene overlapping the Gn domain of the polyprotein coding sequence (Supplementary Figure S2). NSm has not been identified among Group IV and V viruses.

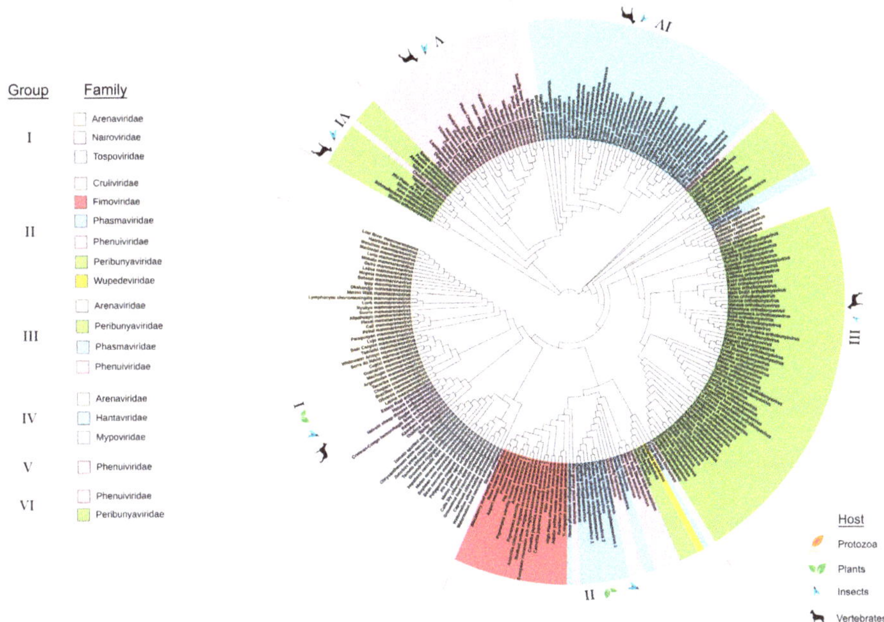

Figure 3. Maximum likelihood phylogenetic tree of the amino acid sequences of the glycoprotein precursor (preGP). Six groups were identified based on clustering from the most distant node. The legend identifies the lineage groups and colors used to identify taxonomic families as in Figure 1. Group I: *Arenaviridae, Nairoviridae* and *Tospoviridae*. Group II: *Cruliviridae, Fimoviridae, Phasmaviridae, Phenuiviridae, Peribunyaviridae,* and *Wupedeviridae*. Group III: *Arenaviridae, Peribunyaviridae, Phasmaviridae,* and *Phenulviridae*. Group IV: *Arenaviridae, Hantaviridae,* and *Mypoviridae*. Group V: *Phenuiviridae*. Group VI: *Phenuiviridae* and *Peribunyaviridae*. Families are color-coded and the hosts for viruses within each group are indicated in the outermost circle. Clade validation is based on the aLRT-SH-like test values.

3.1.3. Phylogeny of Nucleocapsid (N) Proteins

The N proteins of 268 species within the order *Bunyavirales* were used to construct an ML tree with three deeply rooted branches. We identified ten lineage groups (Figure 4) and seven of these groups comprise two or more taxonomic families. Group III contains only *Arenaviridae* and groups IX and X contain only *Phenuiviridae*. Group IX and X include vertebrate and insect-infecting members of *Phenuiviridae*. One similarity between the N and RdRp phylogenies is that the CHIV clusters with the *Pacuvirus* and *Herbevirus* genera in lineage group I along with the family *Peribunyaviridae* (Supplementary Figure S1). Locating CHIV in group I suggests that its ICTV taxonomic classification may be erroneous [42].

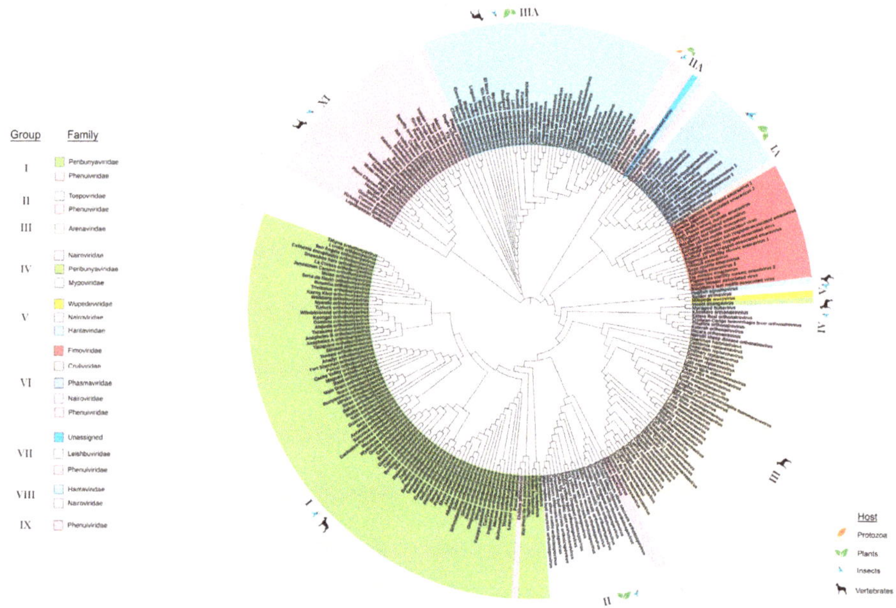

Figure 4. Maximum likelihood phylogenetic tree of the amino acid sequences of the nucleocapsid (N). Ten lineage groups were identified based on clustering from the most distant node. Group I: *Peribunyavirdae* and *Phenuiviridae*. Group II: *Tospoviridae* and *Phenuiviridae*. Group III: *Arenaviridae*. Group IV: *Nairoviridae*, *Peribunyaviridae*, and *Mypoviridae*. Group V: *Wupedeviridae*, *Nairoviridae*, and *Hantaviridae*. Group VI: *Fimoviridae*, *Cruliviridae*, *Phasmaviridae*, *Nairoviridae*, and *Phenuiviridae*. Group VII: *Leishbuviridae*, *Phenuiviridae*, and unassigned species. Group VIII: *Hantaviridae* and *Nairoviridae*. Group IX: *Phenuiviridae*. Group X: *Phenuiviridae*. Families are color-coded and the hosts for viruses within each group are indicated in the outermost circle. Clade validation is based on the aLRT-SH-like test values.

One deeply rooted branch leads to lineage group VII and subsequent subclades arising from this branch include lineage groups I through VI. This large clade spanning from groups I to VII includes the families *Peribunyaviridae*, *Phasmaviridae*, *Fimoviridae*, *Phenuiviridae*, *Tospoviridae*, *Arenaviridae*, and *Nairoviridae* (Figure 4). This deep branch leads directly to Group VII viruses which include the unassigned *Coguvirus*, *Leishbuviridae*, and *Phenuiviridae* members that infect protozoa, arthropods and plants. Each subclade includes an arthropod-infecting genus except for group III Arenaviridae which are vertebrate infecting viruses. For example, the *Herbevirus* genus of group I viruses infects mosquitoes. There are two insect-infecting members of the *Phenuiviridae* in group II that likely gave rise to *Tospoviridae*. Groups IV, V, VI, and VII have the deepest branches associated with protozoan or arthropod infecting viruses. A large component of group VIII includes *Hantaviridae*, insect and plant-infecting *Phenuiviridae*, and one *Nairoviridae* member.

3.2. Phylogeny of MP of Plant Virus Genera Orthotospovirus, Emaravirus and Tenuivirus

Plant virus genomes encode MPs that facilitate intercellular movement and long-distance movement through the vasculature. Researchers identified the *Emaravirus* RNA4 that encodes the 42 kDa P4 protein [43,44], the *Tenuivirus* NS4 [12,45,46], and the *Orthotospovirus* NSm protein as the viral MPs. Previous sequence and structural analysis determined these proteins affiliate with the '30K superfamily' of viral MPs which contain a conserved core of mostly beta-strands [47]. Pairwise comparisons of 42 MP sequences showed most species within the *Emaravirus*, *Tenuivirus*, or *Orthotospovirus* genera shared 60–100% identity and had fewer common residues between the genera (Figure 5). *Emaravirus* MPs formed three subgroups (Figure 4). The first subgroup shares more than 75% identity and includes the species *Ti ringspot associated virus*, *Palo verde broom virus*, *Jujube yellow mottle associated virus*, and *Raspberry leaf blotch virus*. The second subgroup includes *Camellia japonica associated viruses* 1 and 2, and *High Plains wheat mosaic virus*. The third subgroup includes 11 species that share 60% or more identity: *Actinidia chlorotic ringspot-associated virus*, *Redbud yellow ringspot-associated virus*, *Actinida virus 2*, *Pigeonpea sterility mosaic virus 1* and 2, *Fig mosaic virus*, *Pistacia virus*, *Aspen mosaic associated virus*, *Rose rosette virus*, *Blackberry leaf mottle-associated virus*, and *European mountain ash ringspot-associated virus*. Among tenuiviruses, the *Rice grassy stunt virus* shared less than 50% identity with other genus members. There were two groups of orthobunyaviruses that shared more than 80% identical residues (Figure 5).

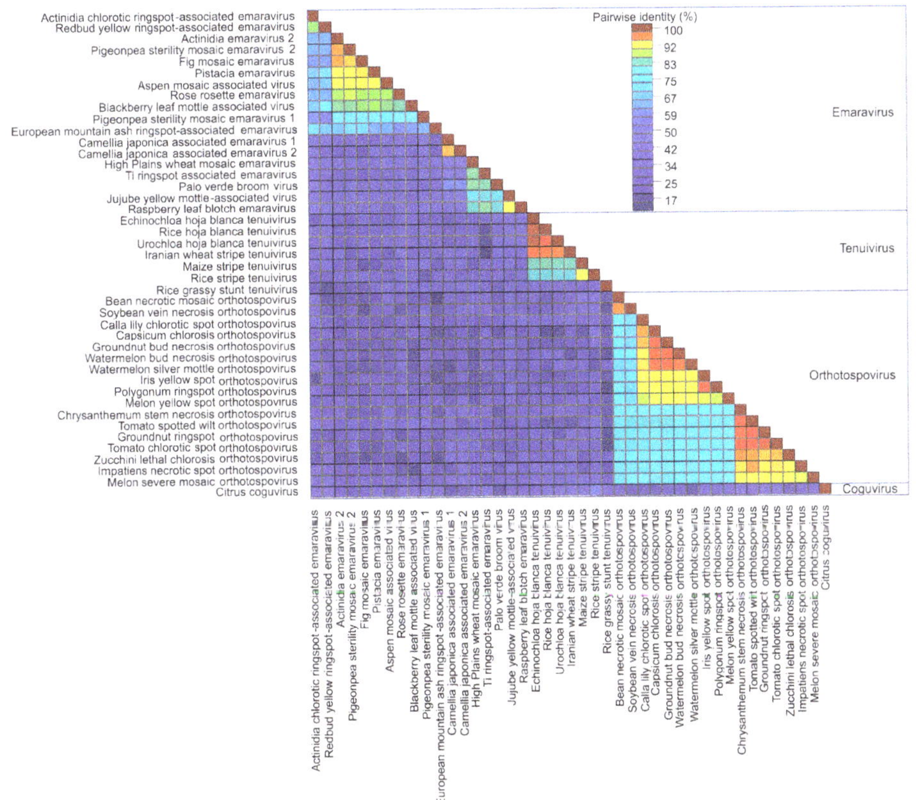

Figure 5. Pairwise sequence alignment of movement proteins (MP) for plant viruses within *Bunyavirales*. Sequence analysis was conducted for all available plant virus within *Bunyavirales*. The plant virus families are *Emaravirus*, *Tenuivirus*, *Orthotospovirus*, and *Coguvirus*. The alignment is colored based on the sequence similarity.

An ML tree showed the MPs in three major clades. Group I consists of the *Fimoviridae* and *Coguvirus* MPs. Group II contains the *Tospoviridae* as well as the *Rice grassy stunt tenuivirus* MPs. Group III is comprised of MPs belonging to *Phenuiviridae* (Figure 6). Conserved structural features of viral movement proteins within the 30K superfamily have been well studied [44,47,48]. Given the number of newly identified species of plant-infecting viruses of *Fimoviridae*, *Phenuiviridae*, and *Tospoviridae*, the multiple sequence alignment shows a low percentage of conserved residues (~18%) across all families (Supplementary Figure S3). Since there is a prevalence of hydrophobic residues (Φ) across the sequences, we manually reviewed the alignment to look for obvious patterns. Notably, all 30K superfamily members have a conserved aspartic acid (D) residue that is found in these 42 movement proteins and is referred to as the "D motif" [47,48]. We determined that the emaraviruses and orthotospoviruses have a common motif surrounding the D motif: Φ-X-Φ-P-X$_{(14)}$-D-X$_{(52-63)}$-W, while the tenuiviruses have a submotif Φ-X-Φ-P-D. The W residue is not conserved downstream of the D motif in the tenuivirus MPs (Supplementary Figure S3).

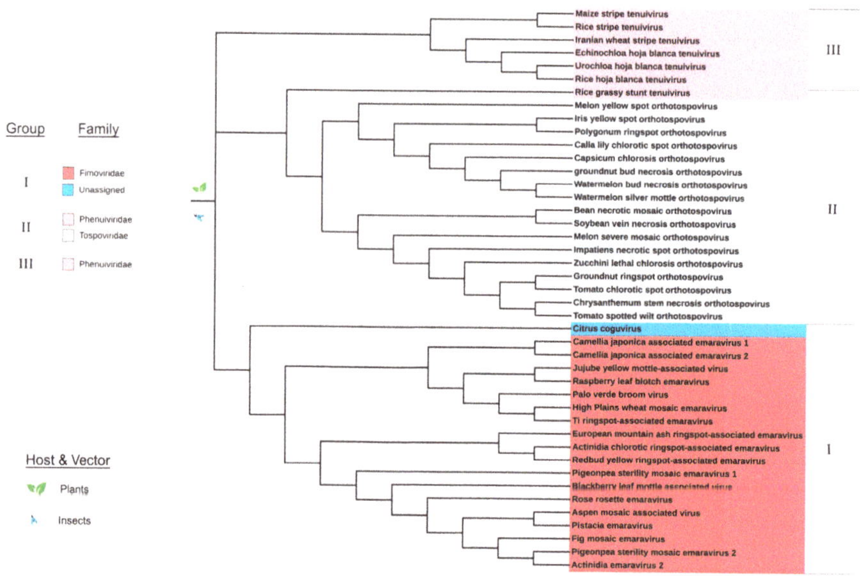

Figure 6. Maximum likelihood phylogenetic tree of the amino acid sequences of the movement protein (MP) belonging to plant viruses. Three groups were identified based on clustering from the most distant node: Group I: *Fimoviridae* and an unassigned species. Group II: *Phenuiviridae* and *Tospoviridae*. Group III: *Phenuiviridae*. All viruses of *Bunyavirales* with an available MP sequence have plant hosts. Clade validation is based on the aLRT-SH-like test values.

3.3. Common Features of Complementary 3′ and 5′ Terminal Regions of Genome Segments

The coding regions of each genome segment lie between terminal non-translated sequences that vary in length. The 3′ and 5′ genomic RNA termini are essential for RNA synthesis and are typically invariant. We compiled the terminal 20 nucleotides for all species that were used in the phylogeny into a table, leaving gaps for those whose sequences were not reported (Supplementary Table S2). We then trimmed the sequences to the first six nucleotides (Supplementary Table S3) and determined these are largely identical within each genus. Sequence logos were created for each family and there was a remarkable level of sequence identity within virus families (Figure 7). The most striking observation was that the 3′ and 5′ UTRs for *Peribunyaviridae*, *Cruliviridae*, the plant-infecting *Fimoviridae*, and two genera of *Phasmaviridae* (*Feravirus* and *Jonvirus*) had identical terminal sequences. It is interesting to see such conservation among animal, plant, and arthropod-infecting viruses. The species CHIV within the *Phenuiviridae*, which we repeatedly noted to be misclassified phylogenetically with *Peribunyaviridae*, also shares the identical terminal sequences with these virus families. Additionally, the plant-infecting

Phenuiviridae (*Tenuivirus* genus) and *Coguvirus* share identical 5′ ACACAA/G and 3′ U/AUGUGU terminal sequences.

The terminal nucleotides for *Tospoviridae*, *Arenaviridae*, and *Myopviridae* are unique (Figure 7). Notably, the *Orthophasmavirus* differs from *Feravirus* and *Jonvirus* in that they each have mirrored tri-nucleotide repeats but differ by a single conserved nucleotide in each repeat. Where *Feravirus* and *Jonvirus* have 5′ AGU̱AGU̱ and 3′ A̱CUA̱CU, *Orthophasmavirus* has 5′ AGC̱AGC̱ and 3′ G̱CUG̱CU (the unlike nucleotides are underlined). It is also worth noting that there is only one nucleotide difference between the 5′ and 3′ terminal sequences of *Nairoviridae* and *Wupedeviridae*. The *Nairoviridae* has 5′ UCUCA̱A and 3′ UUGAGA while *Wupedeviridae* has UCUCU̱A and U̱A̱GAGA.

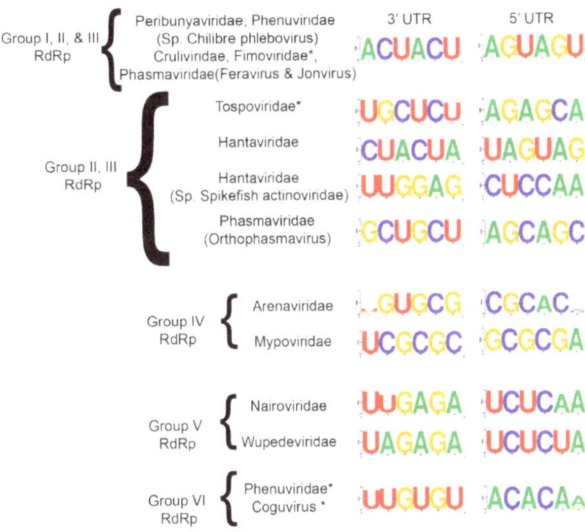

Figure 7. Consensus nucleotide sequence of the 3′ and 5′ termini for each genomic segment of *Bunyavirales*. The consensus sequences were generated using the 6 most distal nucleotides on each end of the viral genomic segments. Each of the analyzed regions was located within a UTR. Families that contain plant viruses are highlighted with an asterisk.

3.4. Cophylogenetic Analysis and Host Range Evolution

Considering the distribution of host taxa on each ML tree, we performed co-phylogeny analysis of virus and host phylogenies at the species level (Figure 8A). These data revealed that duplication and host switching, otherwise known as cross-species transmission, are more common among *Arenaviridae*, *Fimoviridae*, *Hantaviridae*, and *Phasmaviridae* than co-speciation (also known as co-divergence). Duplication is more common than co-speciation or host switching for *Arenaviridae*, *Fimoviridae*, *Nairoviridae*, *Peribunyaviridae*, *Phenuiviridae* and *Tospoviridae*. Considering the preGP, N protein, and MP phylogenies show that vertebrate and plant infecting viruses are related to arthropod infecting viruses suggesting that cross-species transmission may occur between arthropod species, plant species or vertebrate species. However, there is little evidence to suggest the cross-kingdom movement of viruses. The tree also revealed between plant and vertebrate hosts but clustering, host switching during evolutionary history could support the divergent phylogenetic positions for some species within the taxonomic families. Surprisingly, the analysis suggests extinction plays a major role in the evolutionary history for all families in *Bunyavirales* except for *Phasmaviridae* (Figure 8A). The high losses could indicate that there was a mismatch between the independent host and virus phylogenies or descendent of the host species did not inherit a susceptibility to this virus.

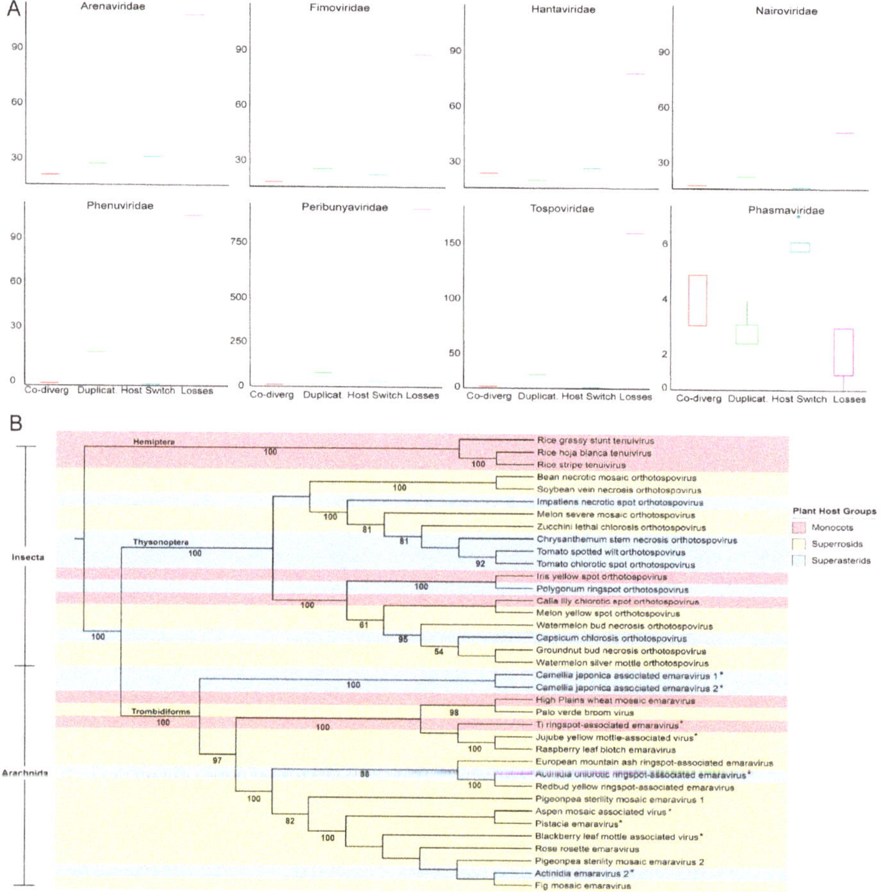

Figure 8. Estimation of co-phylogenetic events of the nucleic acid sequence of plant-infecting virus families within *Bunyavirales*. (**A**). The cophylogeny relationship is based on the RdRp sequences and analyzed using Jane ver. 4.0.1. (**B**). A neighbor joining tree generated using concatenated genomic segments containing RdRp, NC, GP, and MP. Color was used to identify host plant taxonomies and insect vector taxonomy is provided along the branches. Bootstrap values are provided.

To better understand the links between plant infecting viruses, their arthropod vectors, and their plant hosts, an ML tree was generated using concatenated RNA segments representing hallmark genes and MP comprising the genera *Tenuivirus*, *Orthotospovirus*, and *Emaravirus*. Looking at the host spectrum, these plant virus genera are relatively restricted (Figure 8B). Tenuiviruses infect monocot hosts and do not associate with other host types and are transmitted by hemipteran insects. The orthotospoviruses and emaraviruses generally infect members of two large clades of flowering plants known as superrosids and superastrids. Both superrosids and superastrids arose around the same period of rapid evolutionary diversification of eudicots [33,49]. There are two examples of orthotospovirus and emaravirus species infecting monocots. The orthotospoviruses are transmitted by thysonopteran insects and emaraviruses are vectored by trombidiform mites. These plant virus taxa exhibit relatively restricted host and vector spectrum despite the examples of host-switching and low levels of virus-host co-divergence. These data suggest a long-term association between these plant viruses and their hosts although cross-species transmission occurs with some frequency.

A. *Estimation of phylogenetic events within the RdRp.* The amino acid sequence of the RdRp for each family of plant-infecting virus within the order *Bunyavirales* was analyzed and an estimate of co-divergence events (red), duplication events (green), host switch events (blue) and loss events (purple) were summed for each family. Boxes represent the estimated median (center line) interquartile range (IQR) and whiskers represent 1.5× IQR.

B. Maximum-likelihood *tree of viral* segments harboring hallmark genes and movement protein.

A maximum-likelihood phylogenetic tree was constructed from the complete viral genome segments that encode hallmark genes and movement protein (if the sequence is available) for each plant-infecting virus within the order *Bunyavirales*. The genome segments were concatenated in silico before analysis. The virus' vector is listed to the left of the tree, and species with an asterisk (*) have an unknown vector. Each virus is color-code based on its plant host type: monocots (red), superrosids (yellow), or superasterids (blue).

4. Discussion

This study examines the phylogenetic placement of plant viruses within the order *Bunyavirales*. We focused on the genome segments L (or RNA1), M (or RNA2), and S (or RNA3) encoding the RdRp, preGP, and N proteins respectively because they consistently define all members of *Bunyavirales*. We included the analysis of the viral MP because they are a defining feature of plant infecting viruses. This research is timely because, in 2019, the order *Bunyavirales* was amended with significant changes in the associated numbers of families, genera, and species [2]. The ML trees of RdRp, preGP, and N proteins commonly show three deeply rooted branches extending from the base.

The RdRp ML tree shows the plant-infecting *Fimoviridae* and *Tospoviridae* in group II share a common node with *Peribunyaviridae* in group I. While the relatedness of *Tospoviridae* and *Peribunyaviridae* RdRps have been previously reported [6,50], this phylogeny highlights the close relatedness of the *Fimoviridae* and *Tospoviridae* RdRps. The RdRp amino acid sequence alignment shows that the *Orthotospovirus* and *Emaravirus* RdRps share a remarkably high level of conserved residues within the endonuclease and polymerase motifs and that the linear distance between these motifs is similar. These data suggest that selection pressures constrained the amino acid substitutions within these motifs [40].

The RdRp and N form a highly stable complex with viral RNAs that are packaged into virions [40]. The initiation of virus replication requires the formation of a replicative complex that includes the viral RdRp and N proteins. The complementary 3′ and 5′ UTRs of the viral RNA are important for the initiation of replication. The N protein disrupts hydrogen bonding of the "panhandle" structure and enables RNA synthesis by the RdRp [40,51,52]. Given the important engagement between the RdRp, N and UTR regions of the viral RNAs, we expected the N proteins to have similar evolutionary constraints as the RdRp. We were surprised to observe that the N proteins are not as closely related between members of the *Fimoviridae* and *Tospoviridae*. The ML phylogeny of the N proteins showed that the *Tospoviridae* and *Peribunyaviridae* share a common node that bifurcates to groups I and II, while *Fimoviridae* and the arthropod infecting *Phasmaviridae* share a common node in group VI. The complementary 3′ and 5′ termini of the genomic RNA showed a clearer pattern of co-divergence with the lineage groups represented in the RdRp phylogenies. For example, the *Peribunyaviridae*, *Cruliviridae*, *Fimoviridae*, two genera of *Phasmaviridae* (*Feravirus* and *Jonvirus*) and the *Chilibre phlebovirus* share identical terminal 6 nucleotides and the RdRps reside in Groups I, II, and III which derive from a common deep-rooted branch. The *Nairoviridae* and *Wupedeviridae* in Group V RdRp have identical termini except for one nucleotide and the RdRp Group VI *Phenuiviridae* and *Coguvirus* share identical termini. The RdRp Group II and III affiliated *Tospoviridae*, *Hantaviridae*, and *Phasmaviridae* have unique terminal sequences that are shared within these taxonomic families. It is also interesting to point out that the 3′ and 5′ terminal sequences of the plant infecting *Fimoviridae* share identity with the vertebrate infecting *Peribunyaviridae* and not the plant infecting *Tospoviridae*. This observation suggests that the high degree of sequence identity within the RdRp endonuclease and polymerase catalytic motifs of the *Fimoviridae* and *Tospoviridae* is not the driving force for co-evolution of the terminal UTR sequences [53].

However the sequences within neighboring regions of the UTRs that are likely important for replication, transcription, and translation might be influenced by the affinity of the N protein or host factors [38]. For the plant infecting viruses of *Fimoviridae*, *Tospoviridae*, *Phenuiviridae*, and the unassigned *Coguvirus*, experiments are needed to understand how the functional roles of the terminal UTRs.

The CHIV is assigned to the genus *Phlebovirus*; family *Phenuiviridae*. Members of the genus *Phlebovirus* are viruses that are borne by ticks, mosquitoes, and sandflies. Prior molecular characterization suggested that CHIV may be more related to the genus *Pacuvirus* within *Peribunyaviridae* [54]. Our ML analysis using a much larger dataset indicates that CHIV RdRp and N proteins share a specific relationship with the *Pacuvirus* within the Group I *Peribunyaviridae*. The preGP also shows a strong phylogenetic relationship with *Pacuvirus* and *Orthobunyavirus* in group VI. This ML analysis supports the suggestion that the taxonomic identity for CHIV should be moved from the *Phlebovirus* to the *Pacuvirus* genus within *Peribunyaviridae* [54]. Inter-lineage reassortment, although unlikely, may only be considered for the assignment of *Chilibre virus* because these viruses share common reservoirs [55]. However, it is unknown whether an RdRp of one virus can support the replication of such distantly related viruses within this order, arguing against heterotypic reassortment [56–58].

Interestingly, the preGP and N protein phylogenies each exhibited a higher extent of diversity with members of the same taxonomic family sometimes represented in more than one lineage group. For *Phenuiviridae*, the preGP resides in four groups while the N proteins reside in six different groups. Focusing on the plant infecting viruses, the *Tenuivirus* N proteins extend from a deep branch which at its base bifurcates to the insect and vertebrate infecting *Webuvirus*, *Pidchovirus*, and *Orthohantavirus* [59]. The phylogenetic positions of the *Emaravirus* and *Orthophasmavirus* preGP and N proteins suggest a similar ancestry. The *Orthotospovirus* preGP proteins are phylogenetically positioned near the *Orthonairoviruses* and *Mammarenaviruses* while the N proteins extend from a node that is affiliated with two dipteran infecting virus members of *Phenuiviridae* in group II [6]. These observations suggest that the evolutionary connections among viruses with *Bunyavirales* involve a network of gene exchanges. Such gene exchanges likely led to the emergence of new virus species. The data in Supplementary Figure S1 highlight the varying sense and ambisense positions of the preGP open reading frames associated with *Arenaviridae* and *Phenuiviridae* within several phylogenetic groups and strongly suggests the exchange of genes between viruses. An evolutionary mechanism of recombination is supported by the presence or absence of NSm either fused or nested within the preGP coding sequences of neighboring virus genera within a phylogenetic group.

Analysis of the plant viral MPs shows three lineage groups and surprisingly the MPs of *Rice grassy stunt tenuivirus* and orthotospoviruses are closely related in the ML tree. The pairwise analysis also shows that the MP similarities cluster mainly according to the virus genus. All of the MPs in this study have been ascribed to the 30K superfamily of viral MPs which share a common aspartic acid residue that is commonly known as the D motif [47,48]. We identified a larger common motif in the emaraviruses and tospoviruses Φ-X-Φ-P-X$_{(15)}$-D-X$_{(53–59)}$-W, while the tenuiviruses have a submotif Φ-X-Φ-P-D.

Until now horizontal gene transfer among positive-strand RNA viruses and double-strand RNA viruses has been well described but there has been little evidence of gene exchanges among negative-strand RNA viruses [14,60,61]. Horizontal gene exchanges among families within *Bunyavirales* might occur by recombination or reassortment of segments [55,57,62]. Many viruses that infect plants or vertebrates have an insect vector that is responsible for transmission, or that can also serve as an alternative host supporting virus replication. We conducted co-phylogeny analysis to investigate the possibility of segment reassortment or recombination occurring between virus species in common ancestor hosts. Across all families, duplication and host switching were more common than co-divergence with a host species. Looking at the families containing plant viruses, *Fimoviridae* shows duplication, and host switching occurs more than co-divergence while *Tospoviridae* and *Phenuiviridae* show very little host switching. Extinction was high for most families in *Bunyavirales* and this outcome can occur if there is an incongruency between the virus and host phylogenies, when invertebrate vectors narrow the niche diversity, or spill-over infection leads to a dead-end [58,59,63,64]. Extinction may also appear high if the virus and/or host have recently emerged. To better understand the co-phylogeny, we overlaid the plant hosts and insect vector on an ML tree of the plant infecting tenuiviruses, orthotospoviruses, and emaraviruses [65]. The tenuiviruses have four to six genome segments, infect only monocots, and are vectored by hemipteran insects (plant hoppers). The presence of a large segmented genome and its recent origin might have reduced the opportunities for a broader invertebrate vector range. The tospoviruses and emaraviruses infect Superrosids and Superastrids and have clearly separate insect and arachnid vectors. The opportunities for heterotypic reassortment between these genera would more likely occur in a common host than a common vector. Considering the evolutionary history of superrosids and superasterids, these represent two large clades of eudicot plants that emerged approximately 5 million years ago [33,49]. Orthotospoviruses and emaraviruses include a number of recently emerged virus species [43,66,67]. Their emergence may be due to recent commercial trade enabling viruses to move into new geographic regions without expanding the host species diversity. Importantly, the lack of evidence for strong co-speciation argues for a shallow evolutionary clock which may make this study a poor fit for the data.

Our findings provide a comprehensive view of plant virus phylogenetic relationships within the higher ranking of the order *Bunyavirales*. The phylogenies reveal extensive conservation among the hallmark genes of plant-infecting viruses with insect and vertebrate counterparts. The phylogenies reveal important insights into the strength of virus–host and virus–vector interactions. Further research is needed to understand the potential for horizontal gene transfer across diverse virus lineages. A priority for future research is to understand the barriers to virus and host co-speciation that could be critical for preventing epidemic virus spread.

Supplementary Materials: The following are available online at http://www.mdpi.com/1999-4915/12/9/1010/s1, Figure S1: Extracted interior branches of a monophyletic group representing *Chilibre phlebovirus* and neighboring *Pacuvirus* species. Branch support values obtained using the aLRT-SH-like test are provided. All panels show 100% support for monophyletic grouping between *Pacuvirus* and *Chilibre phlebovirus*. A. is the RdRp phylogeny in Figure 1, B. is the preGP phylogeny in Figure 3 and, C. is the N protein phylogeny in Figure 4. Figure S2: Genome structure of M segments from type species of each genera belonging to the order Bunyavirales; Figure S3: Multiple sequence alignment of domains from movement protein (MP) of plant viruses within the order *Bunyavirales*. Table S1: Genome structure, nucleotide, and protein accession numbers of the segments and viral proteins used in this study; Table S2: The 20 distant nucleotides of 5′ and 3′ termini belonging to L, M, and S segments of *Bunyavirales*; Table S3: The 5 distant nucleotides of 5′ and 3′ termini belonging to L, M, and S segments of *Bunyavirales* that used to generate consensus sequences; Table S4: Host species information of the *Bunyavirales* viruses.

Author Contributions: Conceptualization, V.H. and J.V.; methodology, V.H. and J.V.; software, V.H.; validation, V.H., G.R., and J.V.; formal analysis, V.H., G.R.; investigation, V.H. and G.R. and C.D.U.; resources, V.H., J.V.; data curation, V.H., G.R., and C.D.U.; writing—original draft preparation, J.V.; writing—review and editing, J.V., V.H., G.R. and C.D.U.; visualization, V.H., G.R., and J.V.; supervision, J.V.; project administration, J.V.; funding acquisition, J.V. All authors have read and agreed to the published version of the manuscript.

Funding: This research was funded by DBA STAR Roses and Plants, The Conard-Pyle Co. Project # M1902514. This was also partly funded by NSF Project 1759034.

Conflicts of Interest: The authors declare no conflict of interest. The funders had no role in the design of the study; in the collection, analyses, or interpretation of data; in the writing of the manuscript, or in the decision to publish the results.

References

1. Wichgers Schreur, P.J.; Kormelink, R.; Kortekaas, J. Genome packaging of the Bunyavirales. *Curr. Opin. Virol.* **2018**, *33*, 151–155. [CrossRef] [PubMed]
2. Abudurexiti, A.; Adkins, S.; Alioto, D.; Alkhovsky, S.V.; Avšič-Županc, T.; Ballinger, M.J.; Bente, D.A.; Beer, M.; Bergeron, É.; Blair, C.D.; et al. Taxonomy of the order Bunyavirales: Update 2019. *Arch. Virol.* **2019**, *164*, 1949–1965. [CrossRef] [PubMed]

3. Navarro, B.; Minutolo, M.; De Stradis, A.; Palmisano, F.; Alioto, D.; Di Serio, F. The first phlebo-like virus infecting plants: A case study on the adaptation of negative-stranded RNA viruses to new hosts. *Mol. Plant Pathol.* **2018**, *19*, 1075–1089. [CrossRef] [PubMed]

4. Navarro, B.; Zicca, S.; Minutolo, M.; Saponari, M.; Alioto, D.; Di Serio, F. A negative-stranded RNA virus infecting citrus trees: The second member of a new genus within the order bunyavirales. *Front. Microbiol.* **2018**, *9*, 1–11. [CrossRef] [PubMed]

5. Li, C.X.; Shi, M.; Tian, J.H.; Lin, X.D.; Kang, Y.J.; Chen, L.J.; Qin, X.C.; Xu, J.; Holmes, E.C.; Zhang, Y.Z. Unprecedented genomic diversity of RNA viruses in arthropods reveals the ancestry of negative-sense RNA viruses. *Elife* **2015**, *2015*, 1–26. [CrossRef]

6. Guterres, A.; de Oliveira, R.C.; Fernandes, J.; de Lemos, E.R.S.; Schrago, C.G. New bunya-like viruses: Highlighting their relations. *Infect. Genet. Evol.* **2017**, *49*, 164–173. [CrossRef]

7. Ballinger, M.J.; Taylor, D.J. Evolutionary persistence of insect bunyavirus infection despite host acquisition and expression of the viral nucleoprotein gene. *Virus Evol.* **2019**, *5*, 1–9. [CrossRef]

8. Whitfield, A.E.; Huot, O.B.; Martin, K.M.; Kondo, H.; Dietzgen, R.G. Plant rhabdoviruses—Their origins and vector interactions. *Curr. Opin. Virol.* **2018**, *33*, 198–207. [CrossRef]

9. Bojko, J.; Subramaniam, K.; Waltzek, T.B.; Stentiford, G.D.; Behringer, D.C. Genomic and developmental characterisation of a novel bunyavirus infecting the crustacean *Carcinus maenas*. *Sci. Rep.* **2019**, *9*, 1–10. [CrossRef]

10. Käfer, S.; Paraskevopoulou, S.; Zirkel, F.; Wieseke, N.; Donath, A.; Petersen, M.; Jones, T.C.; Liu, S.; Zhou, X.; Middendorf, M.; et al. Re-assessing the diversity of negative strand RNA viruses in insects. *PLoS Pathog.* **2019**, *15*, e1008224. [CrossRef]

11. Koonin, E.V.; Dolja, V.V. Virus World as an Evolutionary Network of Viruses and Capsidless Selfish Elements. *Microbiol. Mol. Biol. Rev.* **2014**, *78*, 278–303. [CrossRef] [PubMed]

12. Kormelink, R.; Garcia, M.L.; Goodin, M.; Sasaya, T.; Haenni, A.L. Negative-strand RNA viruses: The plant-infecting counterparts. *Virus Res.* **2011**, *162*, 184–202. [CrossRef] [PubMed]

13. Simmonds, P.; Adams, M.J.; Benk, M.; Breitbart, M.; Brister, J.R.; Carstens, E.B.; Davison, A.J.; Delwart, E.; Gorbalenya, A.E.; Harrach, B.; et al. Consensus statement: Virus taxonomy in the age of metagenomics. *Nat. Rev. Microbiol.* **2017**, *15*, 161–168. [CrossRef] [PubMed]

14. Wolf, Y.I.; Kazlauskas, D.; Iranzo, J.; Lucía-Sanz, A.; Kuhn, J.H.; Krupovic, M.; Dolja, V.V.; Koonin, E.V. Origins and evolution of the global RNA virome. *mBio* **2018**, *9*, 1–31. [CrossRef]

15. Kuraku, S.; Zmasek, C.M.; Nishimura, O.; Katoh, K. aLeaves facilitates on-demand exploration of metazoan gene family trees on MAFFT sequence alignment server with enhanced interactivity. *Nucleic Acids Res.* **2013**, *41*, 22–28. [CrossRef]

16. Katoh, K.; Rozewicki, J.; Yamada, K.D. MAFFT online service: Multiple sequence alignment, interactive sequence choice and visualization. *Brief. Bioinform.* **2019**, *20*, 1160–1166. [CrossRef]

17. MAFFT Version 7. Available online: https://mafft.cbrc.jp/alignment/server/ (accessed on 12 August 2020).

18. Capella-Gutierrez, S.; Silla-Martinez, J.M.; Gabaldon, T. trimAl: A tool for automated alignment trimming in large-scale phylogenetic analyses. *Bioinformatics* **2009**, *25*, 1972–1973. [CrossRef]

19. Chen, C.; Chen, H.; Zhang, Y.; Thomas, H.R.; Frank, M.H.; He, Y.; Xia, R. TBtools—An integrative toolkit developed for interactive analyses of big biological data. *Mol. Plant* **2020**. [CrossRef]

20. Darriba, D.; Taboada, G.L.; Doallo, R.; Posada, D. ProtTest 3: Fast selection of best-fit models of protein evolution. *Bioinformatics* **2011**. [CrossRef]

21. Guindon, S.; Delsuc, F.; Dufayard, J.-F.; Gascuel, O. Estimating Maximum Likelihood Phylogenies with PhyML. In *Bioinformatics for DNA Sequence Analysis*; Posada, D., Ed.; Humana Press: Totowa, NJ, USA, 2009; pp. 113–137, ISBN 978-1-59745-251-9.

22. Gouy, M.; Guindon, S.; Gascuel, O. SeaView Version 4: A multiplatform graphical user Interface for sequence alignment and phylogenetic tree building. *Mol. Biol. Evol.* **2010**, *27*, 221–224. [CrossRef]

23. Letunic, I.; Bork, P. Interactive Tree Of Life (iTOL) v4: Recent updates and new developments. *Nucleic Acids Res.* **2019**, *47*, W256–W259. [CrossRef] [PubMed]

24. iTOL. Available online: https://itol.embl.de/ (accessed on 9 August 2020).

25. Lorenz, R.; Bernhart, S.H.; Höner zu Siederdissen, C.; Tafer, H.; Flamm, C.; Stadler, P.F.; Hofacker, I.L. ViennaRNA Package 2.0. *Algorithms Mol. Biol.* **2011**, *6*, 26. [CrossRef] [PubMed]

26. Crooks, G.E.; Hon, G.; Chandonia, J.-M.; Brenner, S.E. WebLogo: A sequence logo generator. *Genome Res.* **2004**, *14*, 1188–1190. [CrossRef] [PubMed]

27. WebLogo 3. Available online: http://weblogo.threeplusone.com/ (accessed on 27 April 2020).

28. Muhire, B.M.; Varsani, A.; Martin, D.P. SDT: A virus classification tool based on pairwise sequence alignment and identity calculation. *PLoS ONE* **2014**, *9*, e108277. [CrossRef] [PubMed]

29. Conow, C.; Fielder, D.; Ovadia, Y.; Libeskind-Hadas, R. Jane: A new tool for the cophylogeny reconstruction problem. *Algorithms Mol. Biol.* **2010**, *5*, 1–10. [CrossRef]

30. Taxonomy. Available online: https://www.ncbi.nlm.nih.gov/taxonomy (accessed on 27 July 2020).

31. Virus-Host Database. Available online: https://www.genome.jp/virushostdb/ (accessed on 27 July 2020).

32. Mihara, T.; Nishimura, Y.; Shimizu, Y.; Nishiyama, H.; Yoshikawa, G.; Uehara, H.; Hingamp, P.; Goto, S.; Ogata, H. Linking virus genomes with host taxonomy. *Viruses* **2016**, *8*, 66. [CrossRef]

33. Chase, M.W.; Christenhusz, M.J.M.; Fay, M.F.; Byng, J.W.; Judd, W.S.; Soltis, D.E.; Mabberley, D.J.; Sennikov, A.N.; Soltis, P.S.; Stevens, P.F.; et al. An update of the Angiosperm Phylogeny Group classification for the orders and families of flowering plants: APG IV. *Bot. J. Linn. Soc.* **2016**. [CrossRef]

34. ANGIOSPERM PHYLOGENY WEBSITE, Version 14. Available online: http://www.mobot.org/MOBOT/research/APweb/ (accessed on 11 August 2020).

35. Mielke-Ehret, N.; Mühlbach, H.-P. Emaravirus: A novel genus of multipartite, negative strand RNA plant viruses. *Viruses* **2012**, *4*, 1515–1536. [CrossRef]

36. German, T.L.; Lorenzen, M.D.; Grubbs, N.; Whitfield, A.E. New technologies for studying negative-strand RNA viruses in plant and arthropod hosts. *Mol. Plant Microbe Interact.* **2020**, *33*, 382–393. [CrossRef]

37. Dolja, V.V.; Koonin, E.V. Common origins and host-dependent diversity of plant and animal viromes. *Curr. Opin. Virol.* **2011**, *1*, 322–331. [CrossRef]

38. Amroun, A.; Priet, S.; de Lamballerie, X.; Quérat, G. Bunyaviridae RdRps: Structure, motifs, and RNA synthesis machinery. *Crit. Rev. Microbiol.* **2017**, *43*, 753–778. [CrossRef] [PubMed]

39. Ferron, F.; Weber, F.; de la Torre, J.C.; Reguera, J. Transcription and replication mechanisms of Bunyaviridae and Arenaviridae L proteins. *Virus Res.* **2017**, *234*, 118–134. [CrossRef] [PubMed]

40. Sun, Y.; Li, J.; Gao, G.F.; Tien, P.; Liu, W. Bunyavirales ribonucleoproteins: The viral replication and transcription machinery. *Crit. Rev. Microbiol.* **2018**, *44*, 522–540. [CrossRef]

41. Guardado-Calvo, P.; Rey, F.A. The Envelope Proteins of the Bunyavirales. In *Advances in Virus Research*; Academic Press: Cambridge, MA, USA, 2017; Volume 98, pp. 83–118.

42. Brown, J.K.; Fauquet, C.M.; Briddon, R.W.; Zerbini, M.; Moriones, E.; Navas-Castillo, J. Bunyaviridae. In *Virus Taxonomy*; Elsevier: New York, NY, USA, 2012; pp. 725–741.

43. Ishikawa, K.; Maejima, K.; Komatsu, K.; Netsu, O.; Keima, T.; Shiraishi, T.; Okano, Y.; Hashimoto, M.; Yamaji, Y.; Namba, S. Fig mosaic emaravirus p4 protein is involved in cell to cell movement. *J. Gen. Virol.* **2013**, *94*, 682–686. [CrossRef]

44. Yu, C.; Karlin, D.G.; Lu, Y.; Wright, K.; Chen, J.; MacFarlane, S. Experimental and bioinformatic evidence that raspberry leaf blotch emaravirus P4 is a movement protein of the 30K superfamily. *J. Gen. Virol.* **2013**, *94*, 2117–2128. [CrossRef] [PubMed]

45. Xin, M.; Cao, M.; Liu, W.; Ren, Y.; Zhou, X.; Wang, X. Two negative-strand RNA viruses identified in watermelon represent a novel clade in the order Bunyavirales. *Front. Microbiol.* **2017**, *8*, 1514. [CrossRef]

46. Zhao, W.; Jiang, L.; Feng, Z.; Chen, X.; Huang, Y.; Xue, F.; Huang, C.; Liu, Y.; Li, F.; Liu, Y.; et al. Plasmodesmata targeting and intercellular trafficking of Tomato spotted wilt tospovirus movement protein NSm is independent of its function in HR induction. *J. Gen. Virol.* **2016**, *97*, 1990–1997. [CrossRef]

47. Melcher, U. The "30K" superfamily of viral movement proteins. *J. Gen. Virol.* **2000**, *81*, 257–266. [CrossRef]

48. Mushegian, A.R.; Elena, S.F. Evolution of plant virus movement proteins from the 30K superfamily and of their homologs integrated in plant genomes. *Virology* **2015**, *476*, 304–315. [CrossRef]

49. Moore, M.J.; Soltis, P.S.; Bell, C.D.; Burleigh, J.G.; Soltis, D.E. Phylogenetic analysis of 83 plastid genes further resolves the early diversification of eudicots. *Proc. Natl. Acad. Sci. USA* **2010**, *107*, 4623–4628. [CrossRef]

50. Terret-Welter, Z.; Bonnet, G.; Moury, B.; Gallois, J.-L. Analysis of tomato spotted wilt virus RNA-dependent RNA polymerase adaptative evolution and constrained domains using homology protein structure modelling. *J. Gen. Virol.* **2020**. [CrossRef] [PubMed]

51. Klemm, C.; Reguera, J.; Cusack, S.; Zielecki, F.; Kochs, G.; Weber, F. Systems to establish Bunyavirus genome replication in the absence of transcription. *J. Virol.* **2013**, *87*, 8205–8212. [CrossRef] [PubMed]

52. Guo, Y.; Liu, B.; Ding, Z.; Li, G.; Liu, M.; Zhu, D.; Sun, Y.; Dong, S.; Lou, Z. Distinct mechanism for the formation of the ribonucleoprotein complex of Tomato spotted wilt virus. *J. Virol.* **2017**, *91*, 1–14. [CrossRef] [PubMed]

53. Suzuki, Y.; Kobayashi, Y. Evolution of complementary nucleotides in 5′ and 3′ untranslated regions of influenza A virus genomic segments. *Infect. Genet. Evol.* **2013**, *13*, 175–179. [CrossRef] [PubMed]

54. Hughes, H.R.; Russell, B.J.; Lambert, A.J. Genetic characterization of frijoles and chilibre species complex viruses (genus phlebovirus; Family phenuiviridae) and three unclassified new world phleboviruses. *Am. J. Trop. Med. Hyg.* **2020**, *102*, 359–365. [CrossRef] [PubMed]

55. Briese, T.; Calisher, C.H.; Higgs, S. Viruses of the family Bunyaviridae: Are all available isolates reassortants? *Virology* **2013**, *446*, 207–216. [CrossRef]

56. Wang, J.; Firth, C.; Amos-Ritchie, R.; Davis, S.S.; Yin, H.; Holmes, E.C.; Blasdell, K.R.; Walker, P.J. Evolutionary history of Simbu serogroup orthobunyaviruses in the Australian episystem. *Virology* **2019**, *535*, 32–44. [CrossRef]

57. Roossinck, M.J.; Ali, A. Mechanisms of plant virus evolution and identification of genetic bottlenecks: Impact on disease management. In *Biotechnology and Plant Disease Management*; CABI: Wallingford, UK, 2007; pp. 109–124, ISBN 9781845932886.

58. Shi, M.; Lin, X.D.; Tian, J.H.; Chen, L.J.; Chen, X.; Li, C.X.; Qin, X.C.; Li, J.; Cao, J.P.; Eden, J.S.; et al. Redefining the invertebrate RNA virosphere. *Nature* **2016**, *540*, 539–543. [CrossRef]

59. Holmes, E.C.; Zhang, Y.Z. The evolution and emergence of hantaviruses. *Curr. Opin. Virol.* **2015**, *10*, 27–33. [CrossRef]

60. Koonin, E.V.; Dolja, V.V. Expanding networks of RNA virus evolution. *BMC Biol.* **2012**, *10*, 54. [CrossRef]

61. Liu, H.; Fu, Y.; Jiang, D.; Li, G.; Xie, J.; Cheng, J.; Peng, Y.; Ghabrial, S.A.; Yi, X. Widespread Horizontal Gene Transfer from Double-Stranded RNA Viruses to Eukaryotic Nuclear Genomes. *J. Virol.* **2010**, *84*, 11876–11887. [CrossRef] [PubMed]

62. Webster, C.G.; Reitz, S.R.; Perry, K.L.; Adkins, S. A natural M RNA reassortant arising from two species of plant- and insect-infecting bunyaviruses and comparison of its sequence and biological properties to parental species. *Virology* **2011**, *413*, 216–225. [CrossRef] [PubMed]

63. Geoghegan, J.L.; Senior, A.M.; Di Giallonardo, F.; Holmes, E.C. Virological factors that increase the transmissibility of emerging human viruses. *Proc. Natl. Acad. Sci. USA* **2016**, *113*, 4170–4175. [CrossRef] [PubMed]

64. Geoghegan, J.L.; Duchêne, S.; Holmes, E.C. Comparative analysis estimates the relative frequencies of co-divergence and cross-species transmission within viral families. *PLoS Pathog.* **2017**, *13*, e1006215. [CrossRef] [PubMed]

65. Domingo, E. Viruses at the Edge of Adaptation. *Virology* **2000**, *270*, 251–253. [CrossRef]

66. McGavin, W.J.; Mitchell, C.; Cock, P.J.A.; Wright, K.M.; MacFarlane, S.A. Raspberry leaf blotch virus, a putative new member of the genus Emaravirus, encodes a novel genomic RNA. *J. Gen. Virol.* **2012**, *93*. [CrossRef]

67. Yang, C.; Zhang, S.; Han, T.; Fu, J.; Di Serio, F.; Cao, M. Identification and characterization of a novel emaravirus associated with jujube (*Ziziphus jujuba* Mill.) yellow mottle disease. *Front. Microbiol.* **2019**, *10*, 1–12. [CrossRef]

Perspective

Homo sapiens: The Superspreader of Plant Viral Diseases

Buddhini Ranawaka, Satomi Hayashi *, Peter M. Waterhouse and Felipe F. de Felippes *

Centre for Agriculture and the Bioeconomy, Institute for Future Environments, Queensland University of Technology (QUT), 2 George Street, Brisbane, QLD 4000, Australia; ra.ranawaka@qut.edu.au (B.R.); peter.waterhouse@qut.edu.au (P.M.W.)
* Correspondence: satomi.hayashi@qut.edu.au (S.H.); felipe.felippes@qut.edu.au (F.F.d.F.)

Academic Editors: Michael Goodin and Jeanmarie Verchot
Received: 20 November 2020; Accepted: 15 December 2020; Published: 17 December 2020

Abstract: Plant viruses are commonly vectored by flying or crawling animals, such as aphids and beetles, and cause serious losses in major agricultural and horticultural crops. Controlling virus spread is often achieved by minimizing a crop's exposure to the vector, or by reducing vector numbers with compounds such as insecticides. A major, but less obvious, factor not controlled by these measures is *Homo sapiens*. Here, we discuss the inconvenient truth of how humans have become superspreaders of plant viruses on both a local and a global scale.

Keywords: plant viruses; viral vectors; plant diseases; virus spread

---

1. Introduction

In the year 2020, the world has seen the fast and perverse spread of SARS-CoV-2, which has led to a shutdown of our societies and the loss of 1.3 million human lives worldwide so far [1]. Although unfamiliar to most people, plants are also susceptible to infection by wide range of viruses. Furthermore, damages caused by plant viruses on human lives can be as strong, or even more serious than those caused by their animal counterparts. Throughout history, the outbreak of diseases caused by plant viruses have been major contributors to chronic food insecurity [2], a scenario that tends to worsen with our ever-growing population.

Plant viruses constitute a major cause of plant diseases with an estimated economic impact of more than USD 30 billion annually [3]. Some viruses can wipe out entire plantations, resulting in 100% yield loss [4,5] and, subsequently affecting the revenue of farmers, increasing the price of food, and in more extreme cases, its availability to the market. Globally, the most destructive plant viruses are identified to be members of begomoviruses, tospoviruses and potyviruses. Significant epidemics caused by these viruses include not only those affecting economically important plants, but also staple food crops such as cassava, maize, rice and banana. Therefore, in addition to causing damage to farmers' and countries' economies, such plant disease epidemics can also lead to the starvation of a significant portion of the world's population who depend on these plants for their subsistence [3,6].

While the spread of animal viruses is most often associated with direct contact or proximity to infected individuals, plant viruses are transmitted through wounds on the plant or via a vector, most often insects, fungi and nematodes that feed or infect the plant [3]. Although these vectors have often been the major target for controlling the spread of plant viral diseases, it is apparent that human activities also play a major role in the dissemination of plant viruses (Table 1). Man has distributed most of the cultivated plants around the world by removing them from their centre of domestication. As such, humans are greatly responsible for the novel encounters between plants and their pests [7]. Since many plant viruses have a broad range of hosts and vectors [8], introduction of crops to a new area can enable indigenous viruses from native plants to spread to the crops, and vice versa [3].

Moreover, modern agricultural systems, such as monocultures, have intensified and altered agricultural practices. Continuous cropping patterns encourage the accumulation of viruses and proliferation of their vectors in the field, leading to pandemics.

Table 1. Human intervention in the spread of plant viruses.

Virus	Genus	Vector	Host	Human Intervention
Banana bunchy top virus (BBTV)	Babuvirus	Aphids	Banana, plantain, abaca and other plants in family Musaceae	Trade of vegetative planting and tissue culture material [9]
Banana streak virus (BSV)	Badnavirus	Mealybugs	Banana, *Heliconia*	Trade of vegetative planting and tissue culture material [10]
African cassava mosaic virus (ACMV)	Begomovirus	Whitefly	Cassava, castor bean	Exchange of virus infected plant material. Trade of infected seeds and plant material [11]
Tomato yellow leaf curl virus (TYLCV)	Begomovirus	Whitefly	French bean, Solanaceous plants	Accidental movement of insect vector [12]
Cauliflower mosaic virus (CaMV)	Caulimovirus	Aphids	Cauliflower, Chinese cabbage, brussels sprout, turnip	Virus contaminated machinery, equipment and workers [13].Trade of infected plant material [14]
Cucumber mosaic virus (CMV)	Cucumovirus	Aphids	Soybean, tobacco, pepper	Trade of infected seeds and plant material [15]
Tobacco necrosis virus (TNV)	Necrovirus	*O pidium brassicae*	French bean, cowpea, mung bean, melon, tulip, tobacco, cucumber	Virus contaminated machinery, equipment and workers [16]
Plum pox virus (PPV)	Potyvirus	Aphids	Apricots, peaches, plums, almonds	Grafting, budding, and transplanting of infected plant material [17]
Potato virus Y (PVY)	Potyvirus	Aphids	Potato, tomato, capsicum, tobacco	Virus contaminated machinery, equipment and workers [18]. Trade of infected plant material [19]
Maize dwarf mosaic virus (MDMV)	Potyvirus	Aphids	Maize, sugarcane, sorghum	Trade of infected plant material [20]
Sweet potato feathery mottle virus (SPFMV)	Potyvirus	Aphids	Sweet potato, wild *Ipomoea* sp. *Nicotiana* sp.	Trade of infected tubers and cuttings, grafting, and mechanical inoculation [21]
Zucchini yellow mosaic virus (ZYMV)	Potyvirus	Aphids	Cucumber, pumpkin, rockmelon, zucchini	Virus contaminated machinery, equipment and workers [22]
Sugarcane mosaic virus (SCMV)	Potyvirus	Aphids	Sugarcane and Poaceae plants	Trade of infected plant material [23]
Rice yellow mottle virus (RYMV)	Sobemovirus	Adult beetles in family Chrysomelidae, Grasshoppers	Rice	Virus contaminated sickles, hands and crop residues. Tight contact between plants during planting [24,25]
Tobacco mosaic virus (TMV)	Tobamovirus	Grasshoppers, Caterpillars (occasionally)	Tobacco, tomato, Solanaceous plants	Virus contaminated machinery, equipment, farmers and smokers [26,27]
Tomato spotted wilt virus (TSWV)	Orthotospovirus	Thrips	Peanut, pepper, tomato	Use of infected cuttings and bulbs for propagation [28]
Rice tungro virus (RTV)	Tungrovirus	Leafhopper	Rice	Trade of infected seeds, and transplanting infected seedlings [29]

2. Direct Human Intervention in Virus Spread

There are several ways in which humans currently affect the spread of plant viral diseases. For example, the exchange of virus contaminated material between people plays a major role in transferring the virus to uninfected plants, most often as a result of limited knowledge in viral aetiology of symptomatic plants. For instance, the initial course for the spread of both African cassava mosaic virus (ACMV) [11] and sweet potato virus disease (SPVD) [30] is the exchange of infected stem cuttings and vines, respectively. If farmers are not vigilant, purchasing plant materials (i.e., seeds and tissues for vegetative propagation) from uncertified seed networks can increase the risk of global dissemination of plant virus diseases [31]. The effects are the same with the use of infected plant material for grafting, budding, and transplanting [17].

Another common way in which some viruses spread within crop fields is due to poor agricultural practices, such as the usage of unsterilised tools, not clearing plant debris, and even the continuous use of clothes and shoes that have been in the contaminated field [30,32]. No-till farming is a technique with several benefits to agriculture. However, not removing plant material from one season to another in contaminated fields can spread the virus to new plants and increase its accumulation [30]. Tobacco mosaic virus (TMV) is the typical case where the spread of the disease benefits from continuous cropping system, as it can survive or hibernate in crop debris, soil and other perennial hosts. In addition, these viruses can transmit within the field through mechanical wounds caused by contaminated tools, clothes, and footwear [26]. Interestingly, TMV is also capable of spreading via tobacco products (i.e., air-cured tobacco), where smokers rolling their cigarettes can transmit the virus with their contaminated hands [27].

Some plant diseases rely heavily on insect vectors for the transmission of the virus to a healthy plant. Tomato yellow leaf curl disease is one such case where the disease spreads by the feeding of whitefly vector carrying tomato yellow leaf curl virus (TYLCV) [7]. In this specific example, the insect-mediated viral spread is limited by the flight range of the whiteflies [33]. However, long-distance movement of insect-infested material/commodities by humans have tremendous consequences to how far the insect vector, and therefore the disease, can spread. Indeed, accidental import/export of insect vector-contaminated materials are identified as a major cause of plant virus outbreaks [12].

3. Virus Spread Coupled with Climate Change

The successful emergence and spread of plant viruses, and that of their vectors, are also indirectly influenced by the behaviour of mankind. Global climate change linked to human activities has increased global temperature and CO_2 concentrations, leading to altered rainfall patterns, recurrent extreme weather events, as well as variations in wind velocity and direction [3,34,35]. Such changes have a range of impacts on the host plants, the virus, and their vectors. While some of these events can be beneficial for the plant to fight against infections, an abrupt change in the climatic conditions can also be especially helpful for the dissemination of viral diseases [36]. For example, elevated temperatures have been shown to enhance small RNA mediated defence against ACMV and cymbidium ringspot virus in *Nicotiana benthamiana* [37,38]; however, it also increases the contact transmission, the rate of virus multiplication and systemic movement of the virus within the plant [39]. In addition, higher temperatures are favourable for insects as vectors due to the increase in numbers of winged aphid morphs [40], shorter adult-to-adult generation time [41] and increased flight activity [42]. Moreover, alterations in wind speed and direction can affect how viruliferous vectors disseminate over long distances, affecting their distribution [39].

4. Challenges in Mitigating Plant Viral Diseases

Undoubtedly, lifestyle and reluctance to heed science-based information, at both an individual and societal level, have been major reasons contributing to the current COVID-19 pandemic. Modern people are accustomed to frequent domestic and international travel, large gatherings such as sporting events and concerts, all of which have played a central role in how fast and far the virus has spread. Similarly, the transmission of exotic plant viruses across local and international borders has been aggravated along with increased global trades of food and agriculture products. In addition, food items infected with viruses can easily travel across borders with the world's population travelling more often and further. Overall, trade-in plants, plant products, and the movement of people are accountable for the 71% of factors known as routes of emerging plant viral diseases, while 16% is due to change in the vector populations [43]. A few examples of viruses intercepted at Australian and New Zealand quarantine stations, where strict quarantine measurements are in place, are peanut stripe virus G, apple stem grooving virus, grapevine virus B and sweet potato virus G [43].

The COVID-19 pandemic has shown us the importance of containment measures, such as self-isolation and quarantine, in halting the spread of the disease [44]. The same strategy can also be applied to combat the spread of plant viruses. Indeed, the spread of banana bunchy top virus, potato leafroll virus, sugarcane mosaic virus and plum pox virus have been controlled using effective containment programmes [43] (Figure 1). However, such approaches are limited to situations where there are reliable diagnostics, appropriate infrastructure and community adherence to regulatory protocols. This method is heavily dependent on the commitment and actions of local and federal governments, as well as individuals, which is not always the case. It seems unlikely that the extreme actions leading to changes in our lifestyle, as seen for the COVID-19 crisis, can be easily implemented for fighting against plant viruses.

Figure 1. Containment measures as a strategy to mitigate the spread of plant viral diseases. Some countries such as Australia have strong policies to halt the spread of plant diseases, including viral ones, based on confinement and limitation on the movement of plant material and equipment. Image credit (bottom picture): Biosecurity Queensland.

5. Conclusions

Ultimately, unless the threat of virus infection of food crops is perceived to be of sufficient impact (as may one day be the case due to the escalating world population and reducing areas of fertile arable land), changing human behaviour in order to minimise crop losses seems less likely to be achieved than the development of crops with new sources of virus or vector resistance. To finish on an optimistic note: never before has humanity possessed such extensive genomic information and insights about crops, their wild relatives, their pathogens and their pests; nor has it possessed such powerful molecular and genetic technologies for accelerated breeding and synthetic biology. It is probably with this information and these tools that resilient crops can be developed to increase sustainable food supplies to such a level that they offset the damages wrought by *Homo sapiens*, the superspreader of plant virus diseases.

Author Contributions: B.R., S.H., P.M.W. and F.F.d.F. conceived, designed, and wrote the manuscript. All authors have read and agreed to the published version of the manuscript.

Funding: This work was funded by the Australian Research Council (ARC), grant number FL160100155 and DP170103960.

Acknowledgments: We would like to thank Biosecurity Queensland and Dawit Kidanemariam (QUT) for kindly providing images used in Figure 1.

Conflicts of Interest: The authors declare no conflict of interest.

References

1. COVID-19 Dashboard by the Center for Systems Science and Engineering (CSSE) at Johns Hopkins University (JHU). Available online: https://coronavirus.jhu.edu/map.html (accessed on 18 November 2020).
2. Makkouk, K.M. Plant Pathogens which Threaten Food Security: Viruses of Chickpea and Other Cool Season Legumes in West Asia and North Africa. *Food Secur.* **2020**, *12*, 495–502. [CrossRef]
3. Jones, R.A.; Naidu, R.A. Global dimensions of plant virus diseases: Current status and future perspectives. *Annu. Rev. Virol.* **2019**, *6*, 387–409. [CrossRef] [PubMed]
4. Prasad, A.; Sharma, N.; Hari-Gowthem, G.; Muthamilarasan, M.; Prasad, M. Tomato Yellow Leaf Curl Virus: Impact, Challenges, and Management. *Trends Plant Sci.* **2020**, *25*, 897–911. [CrossRef] [PubMed]
5. Sevik, M.A.; Arli-Sokmen, M. Estimation of the effect of Tomato spotted wilt virus (TSWV) infection on some yield components of tomato. *Phytoparasitica* **2012**, *40*, 87–93. [CrossRef]
6. Rybicki, E.P. A Top Ten list for economically important plant viruses. *Arch. Virol.* **2015**, *160*, 17–20. [CrossRef] [PubMed]
7. Péréfarres, F.; Thierry, M.; Becker, N.; Lefeuvre, P.; Reynaud, B.; Delatte, H.; Lett, J.-M. Biological invasions of geminiviruses: Case study of TYLCV and Bemisia tabaci in Reunion Island. *Viruses* **2012**, *4*, 3665–3688. [CrossRef]
8. Choi, S.; Lee, J.-H.; Kang, W.-H.; Kim, J.; Huy, H.N.; Park, S.-W.; Son, E.-H.; Kwon, J.-K.; Kang, B.-C. Identification of Cucumber mosaic resistance 2 (cmr2) That Confers Resistance to a New Cucumber mosaic virus Isolate P1 (CMV-P1) in Pepper (Capsicum spp.). *Front. Plant Sci.* **2018**, *9*, 1106. [CrossRef]
9. Watanabe, S.; Greenwell, A.M.; Bressan, A. Localization, concentration, and transmission efficiency of Banana bunchy top virus in four asexual lineages of Pentalonia aphids. *Viruses* **2013**, *5*, 758–776. [CrossRef]
10. Lockhart, B. *Diseases of Banana, Abacá and Enset*; Banana Streak Virus: 263–274 (en); Jones, D.R., Ed.; CABI Publishing: Wallingford, UK, 1999.
11. Eni, A.O.; Efekemo, O.P.; Onile-ere, O.A.; Pita, J.S. South West and North Central Nigeria: Assessment of cassava mosaic disease and field status of African cassava mosaic virus and East African cassava mosaic virus. *Ann. Appl. Biol.* **2020**, 1–14. [CrossRef]
12. Caciagli, P. Survival of Whiteflies during Long-Distance Transportation of Agricultural Products and Plants. In *Tomato Yellow Leaf Curl Virus Disease: Management, Molecular Biology, Breeding for Resistance*; Czosnek, H., Ed.; Springer: Dordrecht, The Netherlands, 2007; pp. 57–63.
13. Bak, A.; Emerson, J.B. Cauliflower mosaic virus (CaMV) biology, management, and relevance to GM plant detection for sustainable organic agriculture. *Front. Sustain. Food Syst.* **2020**, *4*, 21. [CrossRef]

14. Cauliflower Mosaic Virus (Cauliflower Mosaic). Available online: https://www.cabi.org/isc/datasheet/15099 (accessed on 25 October 2020).

15. Ali, A.; Kobayashi, M. Seed transmission of Cucumber mosaic virus in pepper. *J. Virol. Methods* **2010**, *163*, 234–237. [CrossRef] [PubMed]

16. Bacsó, R.; Künstler, A.; Király, L. Tobacco necrosis virus replication and spread in Arabidopsis thaliana ecotype Columbia: A potential system for studying plant defense reactions to symptomless virus infections. *Acta Physiol. Plant.* **2016**, *38*, 139. [CrossRef]

17. García, J.A.; Glasa, M.; Cambra, M.; Candresse, T. Plum pox virus and sharka: A model potyvirus and a major disease. *Mol. Plant Pathol.* **2014**, *15*, 226–241. [CrossRef] [PubMed]

18. Sigvald, R. Relationship between aphid occurrence and spread of potato virus Y° (PVY°) in field experiments in southern Sweden. *J. Appl. Entomol.* **1989**, *108*, 35–43. [CrossRef]

19. Potato Virus Y (Potato Mottle). Available online: https://www.cabi.org/ISC/datasheet/43762 (accessed on 25 October 2020).

20. Maize Dwarf Mosaic Virus (Dwarf Mosaic of Maize). Available online: https://www.cabi.org/isc/datasheet/8157 (accessed on 26 October 2020).

21. Maina, S.; Barbetti, M.J.; Edwards, O.R.; de Almeida, L.; Ximenes, A.; Jones, R.A. Sweet potato feathery mottle virus and Sweet potato virus C from East Timorese and Australian Sweetpotato: Biological and molecular properties, and Biosecurity Implications. *Plant Dis.* **2018**, *102*, 589–599. [CrossRef]

22. Gal-On, A. Zucchini yellow mosaic virus: Insect transmission and pathogenicity—The tails of two proteins. *Mol. Plant Pathol.* **2007**, *8*, 139–150. [CrossRef]

23. Srisink, S.; Taylor, P.; Stringer, J.; Teakle, D. An abrasive pad rubbing method for inoculating sugarcane with sugarcane mosaic virus. *Aust. J. Agric. Res.* **1994**, *45*, 625–631. [CrossRef]

24. Kouassi, N.; N'guessan, P.; Albar, L.; Fauquet, C.; Brugidou, C. Distribution and characterization of Rice yellow mottle virus: A threat to African farmers. *Plant Dis.* **2005**, *89*, 124–133. [CrossRef]

25. Abo, M.; Alegbejo, M.; Sy, A.; Misari, S. An overview of the mode of transmission, host plants and methods of detection of rice yellow mottle virus. *J. Sustain. Agric.* **2000**, *17*, 19–36. [CrossRef]

26. McDaniel, L.; Maratos, M.; Farabaugh, J. Infection of plants by tobacco mosaic virus. *Am. Biol. Teach.* **1998**, *60*, 434–439. [CrossRef]

27. Balique, F.; Colson, P.; Raoult, D. Tobacco mosaic virus in cigarettes and saliva of smokers. *J. Clin. Virol.* **2012**, *55*, 374–376. [CrossRef] [PubMed]

28. Department of Agriculture and Fisheries, Queensland Government. Spotted Wilt and Related Viruses. Available online: https://www.daf.qld.gov.au/business-priorities/agriculture/plants/fruit-vegetable/diseases-disorders/spotted-wilt-viruses (accessed on 1 November 2020).

29. Azzam, O.; Chancellor, T.C. The biology, epidemiology, and management of rice tungro disease in Asia. *Plant Dis.* **2002**, *86*, 88–100. [CrossRef] [PubMed]

30. Wokorach, G.; Edema, H.; Echodu, R. Sweetpotato seed exchange systems and knowledge on sweetpotato viral diseases among local farmers in Acholi Sub Region-Northern Uganda. *Afr. J. Agric. Res.* **2018**, *13*, 2551–2562.

31. Tokunaga, H.; Anh, N.H.; Dong, N.V.; Ham, L.H.; Hanh, N.T.; Hung, N.; Ishitani, M.; Tuan, L.N.; Utsumi, Y.; Vu, N.A. An efficient method of propagating cassava plants using aeroponic culture. *J. Crop Improv.* **2020**, *34*, 64–83. [CrossRef]

32. Broadbent, L. Epidemiology and control of tomato mosaic virus. *Annu. Rev. Phytopathol.* **1976**, *14*, 75–96. [CrossRef]

33. Capinera, J. Order Homoptera–aphids, leaf-and planthoppers, psyllids and whiteflies. In *Handbook of Vegetable Pests*, 1st ed.; Academic Press: San Diego, CA, USA, 2001; pp. 279–282.

34. Canto, T.; Aranda, M.A.; Fereres, A. Climate change effects on physiology and population processes of hosts and vectors that influence the spread of hemipteran-borne plant viruses. *Glob. Chang. Biol.* **2009**, *15*, 1884–1894. [CrossRef]

35. Cook, J.; Nuccitelli, D.; Green, S.A.; Richardson, M.; Winkler, B.; Painting, R.; Way, R.; Jacobs, P.; Skuce, A. Quantifying the consensus on anthropogenic global warming in the scientific literature. *Environ. Res. Lett.* **2013**, *8*, 024024. [CrossRef]

36. Krishnareddy, M. Impact of climate change on insect vectors and vector-borne plant viruses and phytoplasma. In *Climate-Resilient Horticulture: Adaptation and Mitigation Strategies*; Springer: Berlin/Heidelberg, Germany, 2013; pp. 255–277.

37. Chellappan, P.; Vanitharani, R.; Ogbe, F.; Fauquet, C.M. Effect of temperature on geminivirus-induced RNA silencing in plants. *Plant Physiol.* **2005**, *138*, 1828–1841. [CrossRef]

38. Szittya, G.; Silhavy, D.; Molnár, A.; Havelda, Z.; Lovas, Á.; Lakatos, L.; Bánfalvi, Z.; Burgyán, J. Low temperature inhibits RNA silencing-mediated defence by the control of siRNA generation. *Embo J.* **2003**, *22*, 633–640. [CrossRef]

39. Jones, R.A.; Barbetti, M.J. Influence of climate change on plant disease infections and epidemics caused by viruses and bacteria. *Plant Sci. Rev.* **2012**, *22*, 1–31. [CrossRef]

40. Diaz, B.M.; Fereres, A. Life table and population parameters of Nasonovia ribisnigri (Homoptera: Aphididae) at different constant temperatures. *Environ. Entomol.* **2005**, *34*, 527–534. [CrossRef]

41. Muñiz, M.; Nombela, G. Differential variation in development of the B-and Q-biotypes of Bemisia tabaci (Homoptera: Aleyrodidae) on sweet pepper at constant temperatures. *Environ. Entomol.* **2001**, *30*, 720–727. [CrossRef]

42. Walters, K.; Dixon, A. The effect of temperature and wind on the flight activity of cereal aphids. *Ann. Appl. Biol.* **1984**, *104*, 17–26. [CrossRef]

43. Rodoni, B. The role of plant biosecurity in preventing and controlling emerging plant virus disease epidemics. *Virus Res.* **2009**, *141*, 150–157. [CrossRef] [PubMed]

44. Bedford, J.; Enria, D.; Giesecke, J.; Heymann, D.L.; Ihekweazu, C.; Kobinger, G.; Lane, H.C.; Memish, Z.; Oh, M.-D.; Schuchat, A. COVID-19: Towards controlling of a pandemic. *Lancet* **2020**, *395*, 1015–1018. [CrossRef]

Article

Development of a New Tomato Torrado Virus-Based Vector Tagged with GFP for Monitoring Virus Movement in Plants

Przemysław Wieczorek *, Marta Budziszewska, Patryk Frąckowiak
and Aleksandra Obrępalska-Stęplowska *

Department of Molecular Biology and Biotechnology, Institute of Plant Protection—National Research Institute,
Władysława Węgorka 20St, 60-318 Poznań, Poland; m.budziszewska@iorpib.poznan.pl (M.B.);
p.frackowiak@iorpib.poznan.pl (P.F.)
* Correspondence: p.wieczorek@iorpib.poznan.pl (P.W.); ao.steplowska@iorpib.poznan.pl (A.O.-S.)

Received: 4 September 2020; Accepted: 19 October 2020; Published: 20 October 2020

Abstract: Green fluorescent protein (GFP)-tagged viruses are basic research tools widely applied in studies concerning molecular determinants of disease during virus infection. Here, we described a new generation of genetically stable infectious clones of tomato torrado virus isolate Kra (ToTV$_{pJL}$-Kra) that could infect *Nicotiana benthamiana* and *Solanum lycopersicum*. Importantly, a modified variant of the viral RNA2—with inserted sGFP (forming, together with virus RNA1, into ToTV$_{pJL}$-Kra$_{GFP}$)—was engineered as well. RNA2 of ToTV$_{pJL}$-Kra$_{GFP}$ was modified by introducing an additional open reading frame (ORF) of sGFP flanked with an amino acid-coding sequence corresponding to the putative virus protease recognition site. Our further analysis revealed that sGFP-tagged ToTV-Kra was successfully passaged by mechanical inoculation and spread systemically in plants. Therefore, the clone might be applied in studying the in vivo cellular, tissue, and organ-level localization of ToTV during infection. By performing whole-plant imaging, followed by fluorescence and confocal microscopy, the presence of the ToTV$_{pJL}$-Kra$_{GFP}$-derived fluorescence signal was confirmed in infected plants. All this information was verified by sGFP-specific immunoprecipitation and western blot analysis. The molecular biology of the torradovirus-plant interaction is still poorly characterized; therefore, the results obtained here opened up new possibilities for further research. The application of sGFP-tagged virus infectious clones and their development method can be used for analyzing plant virus interactions in a wide context of plant pathology.

Keywords: tomato torrado virus; sGFP; plant pathology; infectious clone; plant-virus interaction

1. Introduction

In vivo monitoring of virus movement, as well as its subcellular localization in an infected host plant, can answer many important questions at the forefront of modern molecular plant virology. Currently, it is possible to characterize virus localization in tissues using a wide range of monitoring approaches (reviewed in [1]). Nevertheless, it is beneficial to use in vivo monitoring techniques based on labeling the target virus with a small nontoxic and easily detectable marker. Currently, green fluorescent protein (GFP) seems to be the gold standard in studies focusing on expressing foreign genes from virus genomes [2]. For this purpose, a wide range of virus-based expression systems were developed and applied either in plants [3–5] or animal systems [6].

Tomato torrado virus (ToTV) is a type member of the *Torradovirus* genus within the *Secoviridae* family. ToTV efficiently infects *Solanum lycopersicum*, inducing severe necrosis in tomato and resulting in plant crop loss. Other plant species were also previously described as hosts for ToTV [7], including *Nicotiana benthamiana*, a well-characterized model organism widely used in plant-pathogen interaction studies [8]. Tomato torrado virus is transmitted by whiteflies: *Trialeurodes vaporariorum*, *T. abutilonea* and *Bemisia tabaci* [9]. ToTV, similar to other torradoviruses, has a bipartite single-stranded genome and consists of RNA1 (7829 nt) and RNA2 (5404 nt), tailed with a poly-A track [10]. The self-replicating RNA1 of torradoviruses [11] encodes a single long polyprotein with protease cofactor (ProCo), helicase (Hel), viral protease (Pro), and RNA-dependent RNA polymerase (RdRP) motifs. RNA2 encodes ORF1 (necessary for systemic infection [12]) overlapping the long ORF2 encoding movement protein (3A) followed by three capsid protein (CP) subunits (Vp35, Vp26, and Vp23). Recently, several pathogenicity determinants were described for ToTV [13–15] and tomato marchitez virus (ToMarV) [12].

In this study, we designated ToTV$_{\text{pJL}}$-Kra, the second generation of infectious clones of ToTV (isolate Kra). In these novel clones, cDNA copies of ToTV genomic RNAs were cloned between the 35S promoter of cauliflower mosaic virus (CaMV) and hepatitis delta virus (HDV) ribozyme followed by the 35S terminator of CaMV in a low-copy pJL89 plasmid vector, which is widely used in engineering infectious clones of plant viruses [11,16,17]. Moreover, the sGFP-tagged version of ToTV RNA2 was developed, and together with ToTV RNA1, composed ToTV$_{\text{pJL}}$-Kra$_{\text{GFP}}$, efficiently infected *N. benthamiana* and tomato plants. During ToTV$_{\text{pJL}}$-Kra$_{\text{GFP}}$ infection, sGFP was produced from the RNA2-encoded polyprotein, which was confirmed by monitoring green fluorescence in plants (verified by fluorescence and confocal microscopy) and GFP-specific immunoprecipitation (IP) and western blotting. The use of the sGFP-tagged infectious clone allows the monitoring of ToTV transport in the entire plant as well as its localization in infected cells. The sGFP-tagged infectious clone of ToTV, together with the enclosed protocol of production thereof, can be applied for engineering similar constructs for all other members of the *Torradovirus* genus.

2. Materials and Methods

2.1. Plasmid Construction

All the described genetic engineering manipulations were performed according to standard protocols [18]. First, the previously described full genomic copies of infectious clones of ToTV (p35Kra1 and p35Kra2) [19] were subcloned from their backbone vector (pGreen) to the destination vector pJL89 (the pJL89 plasmid was kindly provided by Professor Masimo Turina). Briefly, 20 ng of the pJL89 plasmid was PCR-amplified with CloneAmp HiFi PCR Premix (Takara, Kusatsu, Shiga, Japan) and the primer pair pJL89ToT1_R/pJL89ToT1i2_F or pJL89ToT2_R/pJL89ToT1i2_F to generate pJL89-based plasmids suitable for cloning full cDNA copies of ToTV RNA1 or RNA2, respectively (Table 1). Next, the full-length cDNA copies of the ToTV RNAs were PCR-amplified using 20 ng of template plasmids (either p35Kra1 or p35Kra2) with CloneAmp HiFi PCR Premix (Takara) and the primers asTo1A_pJL_FW/asTo2C_pJL_RV or asTo2A_pJL_FW/asTo2C_pJL_RV for RNA1 or RNA2, respectively.

One hundred nanograms of the PCR product of the plasmid backbone was mixed with 100 ng of amplified virus cDNA in the presence of 1× NEBuilder HiFi DNA Assembly Master Mix (NEB, Ipswich, MA, USA). The resulting mixture was *DpnI* treated (Thermo Scientific, Waltham, MA, USA) and transformed into *E. coli* Stellar competent cells (Takara). The resulting plasmids pJL89-Kra1 and pJL89-Kra2 were isolated from *E. coli* transformants, sequenced, and tested for their infectivity in *N. benthamiana* and *S. lycopersicum* (Betalux cultivar) (as described previously [19] and further in the paragraph). Next, pJL89-Kra2 was used as a backbone for preparing the RNA2-based expression vector. For this purpose, the additional sequence encoding the putative protease recognition site flanking the C$_{2132}$AG/GTG$_{2137}$ codons (encoding the Q481/V482 putative protease cleavage site between the 3A and Vp35 motifs within the long polyprotein encoded by ToTV RNA2 [20]) was introduced within pJL89-Kra2 using the primers pJLRNA2_CASF4 and pJLRNA_CASR4 and utilizing CloneAmp HiFi PCR Premix (Takara). Next, the coding sequence of sGFP was PCR amplified using the primers EGFP_CASF3 and EGFP_CASR3, and the resulting cDNA was inserted within pJL89-Kra2 using the Gibson assembly protocol. The created construct pJL89-Kra2-GFP was Sanger sequenced and used for the transformation of *Agrobacterium tumefaciens* GV3101.

Table 1. Primers used in this study.

Primer Name	Sequence 5′→3′	Purpose
pJL89ToT1_R	TATATTCTCAAAATAACTCTTTTAA CCTCTCCAAATGAAATGAACTTCC	Amplification of plasmid backbone for cloning of cDNA copy of the ToTV-Kra RNA1
pJL89ToT1i2_F	TTTAAAAAAAAAAAAAAAAAAAAA AAAAAGGGTCGGCATGGCATCTC	
pJL89ToT2_R	TATTGTATAAAATTATTCTTTTAAAC CTCTCCAAATGAAATGAACTTCC	Amplification of plasmid backbone for cloning of cDNA copy of the ToTV-Kra RNA2
pJL89ToT1i2_F	TTTAAAAAAAAAAAAAAAAAAAAA AAAAAGGGTCGGCATGGCATCTC	
asTo1A_pJL_FW	TTTCATTTGGAGAGGTTAAAAGAG TTATTTTGAGAATATAAC	Amplification of cDNA copy of ToTV RNA1
asTo2C_pJL_RV	ATGCCATGCCGACCCTTTTTTTTTT TTTTTTTTTTTTTTTAAAAT	
asTo2A_pJL_FW	TTTCATTTGGAGAGGTTTAAAAG AATAATTTTATACAATATTTATGT	Amplification of cDNA copy of ToTV RNA2
asTo2C_pJL_RV	ATGCCATGCCGACCCTTTTTTTTTT TTTTTTTTTTTTTTTAAAAT	
pJLRNA2_CASF4	AGCAAGACAGCGCTTCTATAAAG AAGCAGCGAAAGCGCAAGTGAA AAACAAGGTGGCCCAAAC	Preparation ToTV-Kra copy of RNA2 variant with duplicated protease recognition site
pJLRNA_CASR4	GGAATCTCCTCGACGGAGGTCTGC GCTACTTTATTCTTAACCTGAGCC TTGGCCGCCTC	
EGFP_CASF3	ACCTCCGTCGAGGAGATTCCGTCAAC GAGCTTTGCGACCATGGTGAGCA AGGGCGAG	Amplification of sGFP coding sequence
EGFP_CASR3	TATAGAAGCGCTGTCTTGCTTGCTCTT TCGCTACGCGTTCCTTGTACAGCT CGTCCAT	
2TT5	GATGAGAAAGGAAAGAAGCAG	Reverse transcription-polymerase chain reaction (RT-PCR) for detection of ToTV variants in plants
2TT6	CATATCACCCAAATGCTTCTC	
3A/Vp35seqF	CCCTTTGATTGTTATGATGGCTT	
3A/Vp35seqR	TGGGCCTTACAGCTTCATTG	
GFP_F	ATGGTGAGCAAGGGCGAG	
GFP_R	CTTGTACAGCTCGTCC	

2.2. Plant Material, Agroinfiltration, and Sap Inoculation

Nicotiana benthamiana and tomato (*S. lycopersicum*, Betalux cultivar) seeds were germinated in an autoclaved universal growth medium in pots in a growth chamber. When the first two true leaves were fully expanded, the seedlings were transplanted individually into 98-cell seed germination trays containing the same universal growth medium and were watered daily. After 10 days of growing, the seedlings were transplanted individually into plastic pots (10 cm in diameter) for further growth and were maintained in a greenhouse with a photoperiod and temperature of 16 h 28 °C/8 h 24 °C (day/night).

Agroinfiltration was performed as described previously [19]. Briefly, a single colony of the recombinant *A. tumefaciens* bacteria was grown in liquid LB medium (supplemented with 50 mg/L rifampicin and 100 mg/L kanamycin) at 28 °C for 48 h with shaking. Afterward, the bacteria were pelleted by centrifugation and resuspended to an OD600 = 1.0 in agroinfiltration buffer (10 mM 2-(N-morpholino)ethanesulfonic acid pH 5.6, 10 mM $MgCl_2$, 200 µM acetosyringone).

Sap inoculation was performed as described by Budziszewska et al. [21]. Briefly, plant material (collected from the 3 infected plants) was ground with a mortar and pestle in the presence of 0.1 M phosphate buffer and mechanically inoculated onto carborundum-dusted leaves of tested plants.

2.3. Virus Detection by RT-PCR

Virus detection in systemic leaves of infected/infiltrated plants was performed utilizing RT-PCR. To achieve this aim, total RNA was isolated using TriReagent (Thermo Scientific) and precipitated with isopropanol [22]. The resulting total RNA (ca. 1 µg) was converted to cDNA using 200 U of RevertAid Reverse Transcriptase (Thermo Scientific) in the presence of 50 ng of random hexamers (Thermo Scientific). For RT-PCR, 1 µL of cDNA was used in a 20 µL reaction containing 1× DreamTaq PCR Master Mix (Thermo Scientific) in the presence of a 500 nM mixture of forward and reverse primers (Table 1).

2.4. Fluorescence Monitoring in Plants

Fluorescence was monitored in whole plants using a VersaDoc 4000 MP Imaging System (Bio-Rad, Hercules, CA, USA) set with the following parameters: Light mode: LED epi, color: Blue, filter name: 530 BP, gain setting: 1×, and exposure time: 3–120 s.

For microscopy analysis, 2 leaf disks were mounted in water between a slide and cover glass with the upper epidermis forward with a 10× objective. Fluorescence microscopy analysis was performed using a BX53 microscope (Olympus, Shinjuku, Tokyo, Japan) with a GFP-specific filter. Laser-scanning confocal microscopy was performed at the Laboratory of Electron and Confocal Microscopy (Faculty of Biology, Adam Mickiewicz University, Poznań, Poland).

Additionally, fluorescence was measured in a crude extract prepared from plants verified for fluorescence by a DTX 880 Multimode Detector (Beckman Coulter, Brea, CA, USA) For this experiment, 3 disks (5 mm in diameter) from each leaf were sampled and homogenized in 100 µL of sterile water, followed by centrifugation to remove the plant debris. The resulting supernatant was taken for analyses. Fluorescence was measured in a black 96-well plate with a clear bottom using 485/535 nm (excitation/emission) filters.

2.5. Immunodetection (IP) of the Recombined GFP Protein

Recombinant GFP was pulled down from $ToTV_{pJL}$-Kra_{GFP}-infected plants using IP using GFP-Trap Magnetic Agarose (Chromotek, Planegg-Martinsried, Germany). Briefly, plant material (250–500 mg) was pulverized in liquid nitrogen, followed by homogenization in RIPA buffer (Thermo Scientific). The homogenate was mixed by vortexing for 1 min, followed by centrifugation (14000 rpm for 10 min at 4 °C) to remove plant debris. The resulting supernatant was used for IP.

The GFP-Trap Magnetic Agarose (Chromotek) was gently resuspended (25 µL of the bead slurry per sample) and equilibrated in ice-cold RIPA buffer (Thermo Scientific). The beads were separated with a magnet until the supernatant became clear. The equilibration was performed twice. Next, the lysate was added to the equilibrated beads and mixed end over end for 1 h at 4 °C. The beads were separated with a magnet as mentioned above and washed 3 times with 500 µL of ice-cold RIPA buffer. Finally, the beads were boiled in 50 µL of 2× SDS sample buffer for 5 min at 95 °C to dissociate immunocomplexes.

The recombinant GFP was detected by western blotting, as follows: 30 µL of the protein lysate was fractioned by means of sodium dodecyl sulfate-polyacrylamide gel electrophoresis in a 12% polyacrylamide gel followed by protein transfer onto a PVDF membrane. The filter was blocked for 1 h at room temperature with 5% nonfat milk in phosphate-buffered saline with 0.1% Tween (PBS-T) buffer followed by incubation (1 h at room temperature) with a primary antibody (anti-GFP,

Agrisera, Vännäs, Sweden) at a dilution of 1:2000 in blocking buffer. The membrane was washed with PBS-T followed by incubation (1 h at room temperature) with a secondary antibody conjugated with horseradish peroxidase (goat anti-rabbit IgG, Agrisera) at a dilution of 1:10000. After intensive washing, the membrane was developed for 5 min with AgriseraECL SuperBright (Agrisera) solution. Images of the blot were obtained using a CCD imager (VersaDoc 4000 MP Imaging System, Bio-Rad, Hercules, CA, USA).

3. Results

3.1. The New Generation of Infectious Clones of ToTV Retains Their Biological Activity

According to our previous observations, the first generation of infectious clones of ToTV [16] performed well in infectivity assays and was successfully used for analyzing ToTV gene functions in the context of virus pathogenicity [14]. However, the clones were found to maintain low stability during their passages in *E. coli* systems (data not shown). Therefore, we found it essential to improve their genetic stability. By performing subcloning procedures aiming at substituting cDNA copies of RNA1 and RNA2 of ToTV-Kra from its original infectious clones (p35Kra1 and p35Kra2), the second generation of plasmids was obtained: pJL89-Kra1 and pJL89-Kra2. The plasmids were used to transform *A. tumefaciens* GV3101 for subsequent agroinfiltration. By performing infectivity assays, it was demonstrated that the mobilized virus, named hereafter $ToTV_{pJL}$-Kra, was infectious to *N. benthamiana* and *S. lycopersicum*. This was confirmed by disease symptoms manifested on agroinfiltrated plants: yellowing and ToTV-specific spoon-like malformations of systemically infected leaves in *N. benthamiana* and leaf mottling followed by severe necrosis developing near veins of systemically infected leaves in *S. lycopersicum* (Figure 1A, middle panel). The same disease symptoms were observed on plants mechanically inoculated with the infectious sap derived from ToTV-Kra-infected *N. benthamiana* (Figure 1A, left panel). The presence of viral RNAs was confirmed in diseased plants utilizing RT-PCR analysis with primers complementary to ToTV RNA2, resulting in the amplification product of the expected size of 624 bp (Figure 1B).

Since it was confirmed that the novel generation of infectious clones of ToTV could infect plants, the pJL89-Kra2 clone was used for further engineering. By performing Gibson assembly, the pJL89-Kra2-sGFP clone was prepared, in which the sGFP open reading frame (ORF, flanked at the N- and C-ends with an additional sequence encoding putative protease recognition sites) was introduced between the 3A and Vp35 coding sequences (Figure 2). The sGFP ORF was inserted seamlessly between 3A and Vp35, as confirmed by Sanger sequencing of the engineered locus within pJL89-Kra2-sGFP. The clone was used to transform *A. tumefaciens* GV3101 bacteria for subsequent agroinfiltration.

3.2. GFP-Tagged ToTV Infects N. benthamiana, Spreads Efficiently in the Host and Can Be Mechanically Passaged

To test whether pJL89-ToTV-sGFP can infect host plants, 4 to 6-week-old *N. benthamiana* seedlings were agroinfiltrated with a mixture of *A. tumefaciens* harboring the pJL89-Kra1 and pJL89-Kra2-sGFP clones (forming together in a host into $ToTV_{pJL}$-Kra$_{GFP}$). Six days after infiltration, the plant material was collected and checked for the presence of the engineered virus in their systemic leaves. For this, total RNA was extracted from systemic leaves of tested plants, converted to cDNA, and taken for RT-PCR analysis targeting three loci in RNA2: The Vp35/Vp26 ORF (primers 2TT5/2TT6), the engineered junction site 3A/Vp35 (primers seq3A/Vp35_F/seq3A/Vp35_R) and specifically the sGFP ORF (primers GFP_F/GFP_R) (Figure 3A). RT-PCR with the primers 2TT5/2TT6 resulted in amplification products of 624 bp using RNA extracted from both $ToTV_{pJL}$-Kra and $ToTV_{pJL}$-Kra$_{GFP}$-infected plants (Figure 3A).

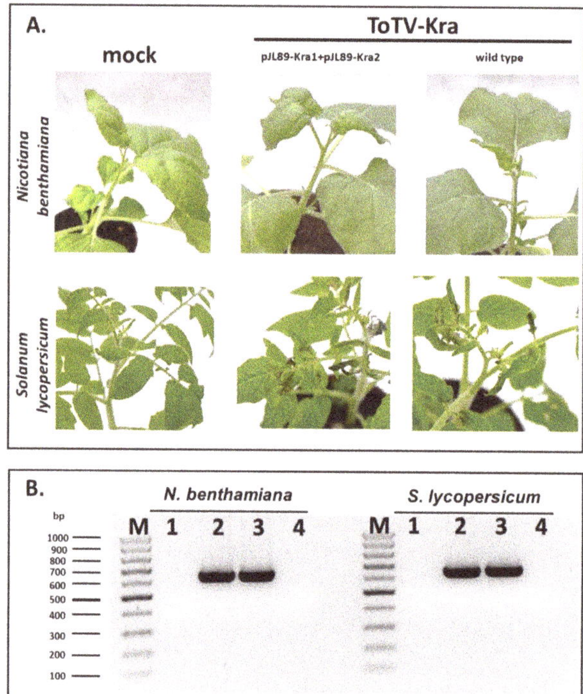

Figure 1. (**A**) Infection symptoms observed in *Nicotiana benthamiana* and *Solanum lycopersicum* (cultivar Betalux) infected with ToTV$_{pJL}$-Kra or wild-type tomato torrado virus (isolate Kra); (**B**) Reverse transcription-polymerase chain reaction (RT-PCR)-based detection of ToTV$_{pJL}$-Kra or wild-type ToTV-Kra in host plants. M-DNA mass ruler, 1—mock-infected plants, 2—ToTV$_{pJL}$-Kra-infected, 3—wild-type ToTV-Kra-infected, 4—no template control; bp- base pairs.

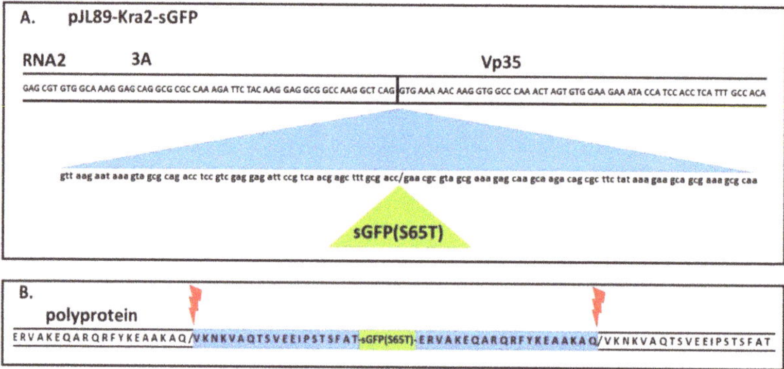

Figure 2. Schematic representation of the modified region of the pJL89-Kra2-GFP infectious clone. (**A**) The putative coding region of the protease recognition site was duplicated (blue) and inserted together with the sGFP (S65T) (green) open reading frame between the 3A/Vp35 junction site; (**B**) The modified virus polyprotein translated from RNA2 with an inserted sGFP open reading frame. The blue region indicates the putative protease recognition site flanking sGFP. The red markers indicate the virus protease cleavage sites at Q/V.

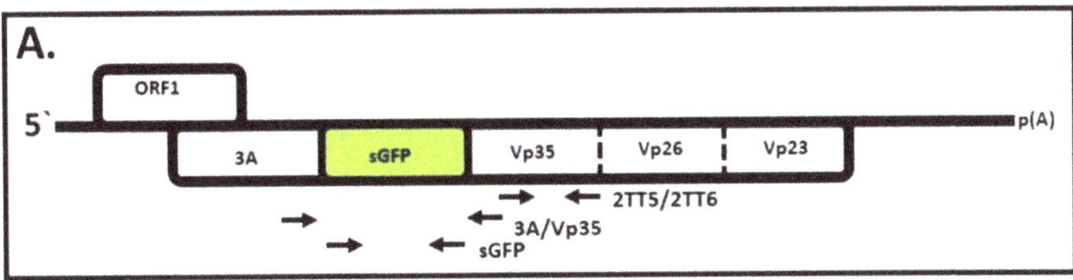

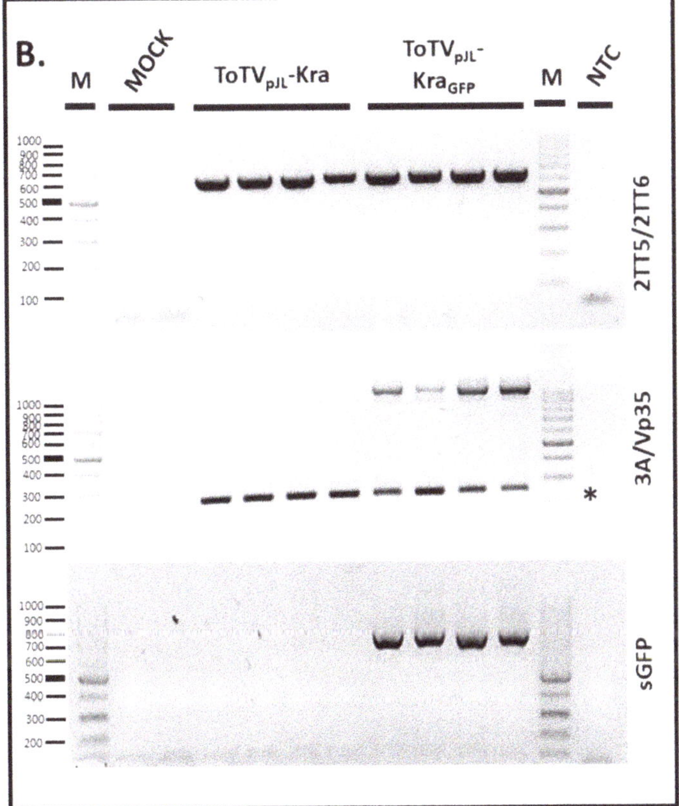

Figure 3. Reverse transcription-polymerase chain reaction (RT-PCR)-based detection of ToTV$_{pJL}$-Kra and ToTV$_{pJL}$-Kra$_{GFP}$ in infected *Nicotiana benthamiana*. Mock-infected plants were also included. (**A**) Schematic representation of the annealing sites of three primer pairs (2TT5/2TT6, 3A/Vp35, and sGFP) used in the RT-PCR detection assays; (**B**) Results of the RT-PCR analysis of the tested plants. The asterisk indicates the amplification products of the locus without sGFP.

In the case of the RT-PCR with primers flanking the 3A/Vp35 engineered region in RNA2, a 249 bp product was expected to be amplified in plants infected by ToTV$_{pJL}$-Kra. Indeed, the amplicon was detected in those plants (Figure 3B). Importantly, the insertion of the sGFP ORF between 3A/Vp35 elongated the tested region by an additional 838 bp. After RT-PCR, the 1087 bp amplification product was detected only in plants infected with ToTV$_{pJL}$-Kra$_{GFP}$. In the same plants, however, an additional RT-PCR product of 249 bp was amplified from the pool of RNA2 lacking the sGFP sequence. Finally, the third RT-PCR performed with the GFP_F/GFP_R primers gave a 717 bp amplification product only in plants infected by ToTV$_{pJL}$-Kra$_{GFP}$ (Figure 3B). All the conducted RT-PCR analyses confirmed that ToTV$_{pJL}$-Kra$_{GFP}$ systemically infected *N. benthamiana*.

Moreover, to test whether ToTV$_{pJL}$-Kra$_{GFP}$ was transmissible from plant to plant, *N. benthamiana* (infected with the modified virus, Supplementary Figure S1) was homogenized, and the obtained sap was used for mechanical inoculation of *N. benthamiana* and *S. lycopersicum* (Betalux cultivar) seedlings. Seven days after inoculation, ToTV$_{pJL}$-Kra$_{GFP}$ was detected in inoculated *N. benthamiana*, as well as in tomato plants (Supplementary Figure S1). These findings were verified by RT-PCR with the aforementioned primer pairs. This result showed that ToTV$_{pJL}$-Kra$_{GFP}$ was infectious and stable through passages in *N. benthamiana* and *S. lycopersicum*.

3.3. GFP-Derived Fluorescence is Detected in Plants Infected with ToTV-GFP

Initially, to verify the expression of sGFP from ToTV$_{pJL}$-Kra$_{GFP}$ in *N. benthamiana*, plants were illuminated under UV using a hand-held lamp. However, under UV light, GFP fluorescence was not detected in ToTV$_{pJL}$-Kra$_{GFP}$-infected plants. Therefore, GFP fluorescence had to be monitored by substantially more sensitive detectors coupled with a CCD camera. Under a blue LED light source and using a 530 BP filter, GFP fluorescence was visualized; within the systemic leaves in plants infected with ToTV$_{pJL}$-Kra$_{GFP}$, fluorescence was monitored and manifested as a strong bright light signal. The signal was detected in veins (vascular tissue) and interveinal areas (mesophyll) of the leaves (Figure 4).

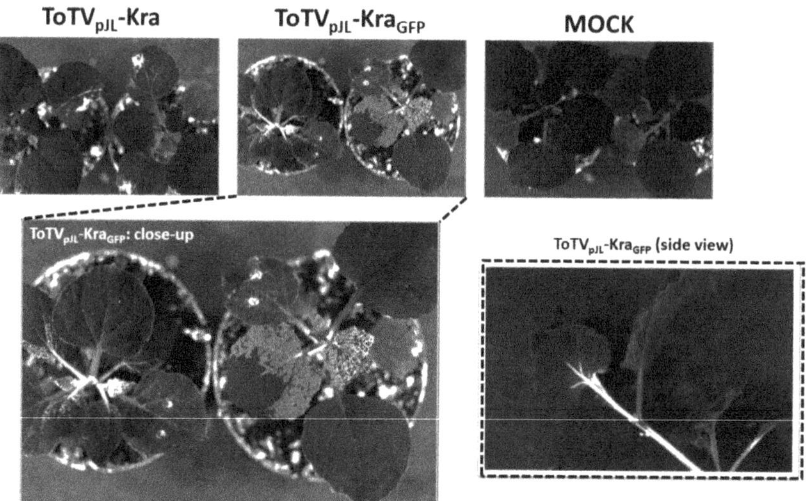

Figure 4. Visualization of sGFP-derived fluorescence in *Nicotiana benthamiana* using a blue LED light source and 530 BP filter. Plants infected with ToTV$_{pJL}$-Kra or ToTV$_{pJL}$-Kra$_{GFP}$ were exposed to a blue LED light source, and fluorescence was detected using a 530 BP filter. Light-gray areas within the systemic leaves of *N. benthamiana* were detected only in plants infected with ToTV$_{pJL}$-Kra$_{GFP}$. The migration of ToTV$_{pJL}$-Kra$_{GFP}$ in the plant (across the main stem, petiole, and primary veins in the leaf) was detected from the side view.

Importantly, using the same detection system, no fluorescence signal was observed in mock- or ToTV$_{pJL}$-Kra-infected plants. To confirm that the bright light signal was derived specifically from the fluorescence of sGFP, the illuminated leaves were analyzed using a fluorescence microscope. In this analysis, the fluorescence signal observed in the cytoplasm and nucleus was detected only in plant material infected by ToTV$_{pJL}$-Kra$_{GFP}$ (Figure 5A).

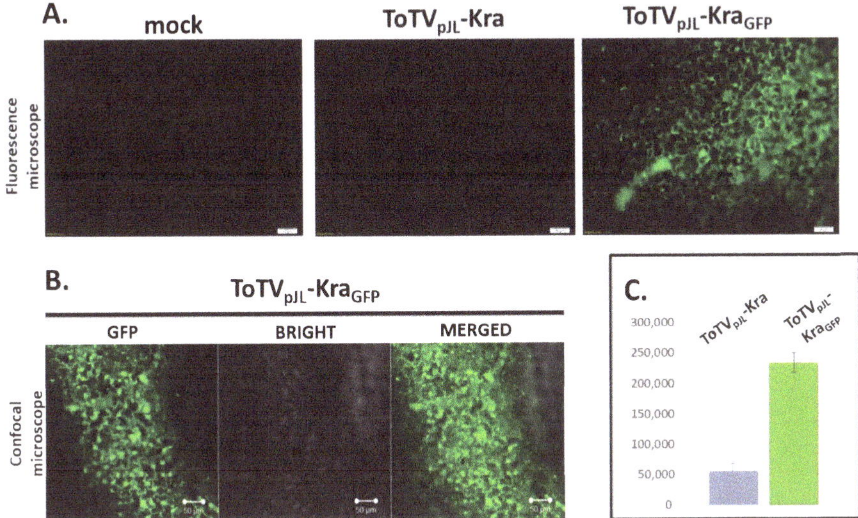

Figure 5. Verification of sGFP-derived fluorescence in *Nicotiana benthamiana* infected with ToTV$_{pJL}$-Kra$_{GFP}$. The analysis was performed using fluorescence (**A**) and confocal microscopy (**B**). Scale bar: Fluorescence microscopy and confocal images = 50 µM; (**C**) Analysis of absolute fluorescence in cleared plant extract of *N. benthamiana* infected by ToTV$_{pJL}$-Kra or ToTV$_{pJL}$-Kra$_{GFP}$.

Additionally, the plants infected by ToTV$_{pJL}$-Kra$_{GFP}$ were subjected to fluorescence detection by confocal microscopy, and again, the fluorescence signal was confirmed in the plant cells (it was also observed from the nucleus) (Figure 5B). Lastly, the fluorescence level was assessed in cleared leaf extract of *N. benthamiana* infected by ToTV$_{pJL}$-Kra or ToTV$_{pJL}$-Kra$_{GFP}$. Again, this result verified substantially elevated fluorescence levels in plants infected with ToTV carrying the sGFP ORF (Figure 5C). In summary, all the performed analyses confirmed that the described ToTV$_{pJL}$-Kra$_{GFP}$ was capable of infecting the plants, which was accompanied by sGFP-derived fluorescence.

3.4. GFP is Produced in N. benthamiana Infected by ToTV-GFP

To finally confirm that sGFP is produced in ToTV$_{pJL}$-Kra$_{GFP}$-infected plants, immunodetection of the heterologous protein was performed using sGFP-specific antibodies. Preliminarily, western blots were performed using a crude plant extract prepared from *N. benthamiana* infected by ToTV$_{pJL}$-Kra$_{GFP}$. However, no sGFP-specific signal was detected in the assay (data not shown). It was assumed that sGFP accumulated at low levels in plants infected by ToTV$_{pJL}$-Kra$_{GFP}$. Therefore, subsequent western blot assays were performed with sGFP IP followed by immunodetection. Indeed, supported by this approach, the mature sGFP (ca. 30 kDa), as well as polyprotein maturation side products (3A-sGFP and sGFP-Vp35, ca. 49 kDa each), were detected only in plants infected by ToTV$_{pJL}$-Kra$_{GFP}$. sGFP was not detected in the ToTV$_{pJL}$-Kra- or mock-infected plants (Figure 6).

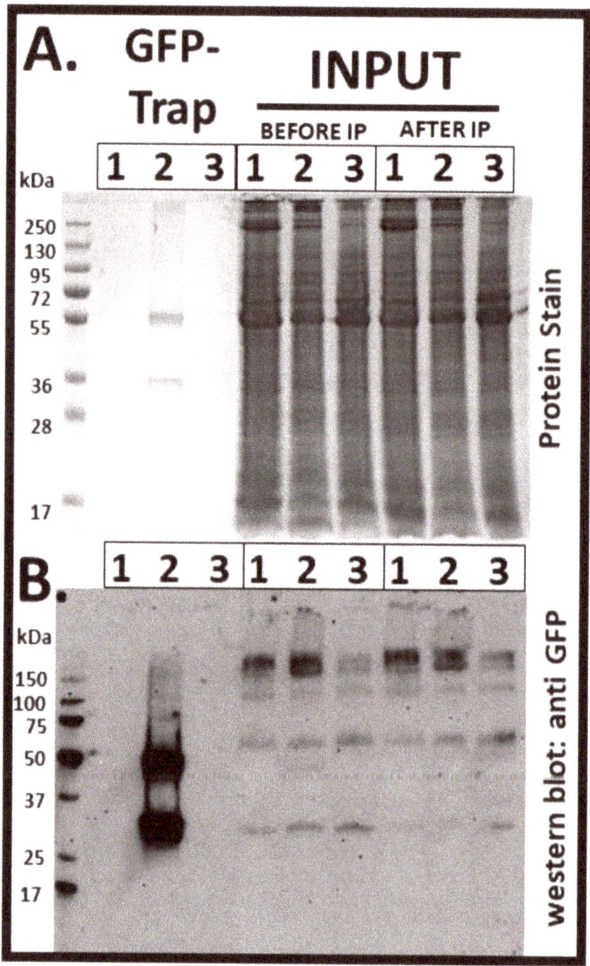

Figure 6. Detection of sGFP accumulation in *Nicotiana benthamiana* infected with ToTV$_{pJL}$-Kra$_{GFP}$ or ToTV$_{pJL}$-Kra. Plant material infected with ToTV$_{pJL}$-Kra (1), ToTV$_{pJL}$-Kra$_{GFP}$ (2), or mock-treated plants (3) were collected and subjected to anti-GFP immunoprecipitation (IP) using GFP-Trap Magnetic Agarose. Protein samples eluted from agarose as well as protein input collected before and after GFP-IP were subjected to sodium dodecyl sulfate-polyacrylamide gel electrophoresis followed by protein staining (**A**) and immunoblot (**B**) analyses. kDa—kilodalton.

4. Discussion

Infectious clones (or infectious transcripts) of plant viruses remain a basic research tool in plant pathology, concerning mostly studies on virus-derived pathogenicity determinants [23]. On the other hand, plant virus vectors were described as platforms for heterologous protein production in plants [24] (revised in [25]) or for silencing gene expression in hosts (most commonly tobacco rattle virus and potato X virus). The goal of this study was to develop stable ToTV-based constructs suitable for expressing a reporter protein, GFP, in plants. During the propagation of the previously described infectious clones p35Kra1 and p35Kra2 [19] in *E. coli*, we found that the constructs could not maintain their stability over time after passaging in the bacteria. This instability manifested with the production of shorter than expected versions of the cloned plasmids after the third round of passaging through *E. coli*. By changing the plasmid backbone (from the original pGreen-based to the pJL89), we observed higher stability of the newly developed infectious clones of the ToTV during their propagation in bacteria. The new generation of infectious clones was able to stably replicate after transferring from low-volume cultures (up to 5 mL) into large-scale ones (up to 300 mL). Indeed, it was described that using low-copy-number plasmids might be beneficial for maintaining the genetic stability of infectious clones [26,27]. Moreover, using the pJL89 plasmid backbone is additionally advantageous because its replication in *A. tumefaciens* does not have to be supported by an additional helper plasmid [28].

In research described by Ferriol et al. [11], a ToMarV isolate M (ToMarV-M, another member of the *Torradovirus* genus) expressing GFP was described. ToMarV-M-GFP was used to verify the self-replicating abilities of RNA1 of the virus and local cell-to-cell movement of the virus. These findings were confirmed by monitoring ToMarV-M-GFP-derived fluorescence, particularly in infiltrated leaves of *N. benthamiana*. In comparison, in our research, we additionally tested the ability for systemic long-distance movement of the recombined virus $ToTV_{pJL}$-Kra_{GFP} within the plant and its ability to infect tobacco and tomato plants via mechanical inoculation. First, we have shown that $ToTV_{pJL}$-Kra_{GFP} can replicate locally and systemically infect *N. benthamiana*. This finding was verified by RT-PCR with sequence-specific primers for ToTV or sGFP. Most importantly, GFP-derived fluorescence, as well as accumulation of the recombinant protein was detected in $ToTV_{pJL}$-Kra_{GFP}-infected plants. However, the recombined sGFP could be specifically isolated after subjecting the same extract to a GFP-specific pull-down assay with GFP-Trap Magnetic Agarose. This result might be explained by the high affinity (dissociation constant KD of 1 pM) and binding capacity (8 µg/10 µl of used suspension) of the used affinity resin. $ToTV_{pJL}$-Kra_{GFP} can be used for tracking virus movement in the host, for instance, in studies concerning host- and virus-derived factors determining the pathogen's host-specific movement [14].

Taken together, all the performed analyses confirmed that heterologous sGFP can be produced in plants using the infectious $ToTV_{pJL}$-Kra_{GFP} clone. More importantly, this clone can be used for monitoring virus cell-to-cell migration as well as long-distance movement in infected plants, which was described here for the first time for the type member of *Torradovirus* genus, in the context of investigating ToTV pathogenicity in the mentioned hosts.

Supplementary Materials: The following are available online at http://www.mdpi.com/1999-4915/12/10/1195/s1, Figure S1: Green fluorescent protein-tagged ToTV (ToTV$_{pJL}$-Kra$_{GFP}$) is infectious and stable through passages in *Nicotiana benthamiana* and *Solanum lycopersicum*.

Author Contributions: Conceptualization, P.W.; designing the idea of the experiment, P.W. and A.O.-S.; methodology, P.W.; performing experiments, P.W., P.F., M.B.; writing—original draft preparation: P.W.; writing—review and editing, A.O.-S., M.B. and P.F.; project administration, P.W. All authors have read and agreed to the published version of the manuscript.

Funding: This research was funded by NATIONAL SCIENCE CENTRE, grant number UMO-2016/21/D/NZ9/02478.

Acknowledgments: We sincerely would like to thank Massimo Turina from the Institute for Sustainable Plant Protection, National Research Council (Turin, Italy) for providing the pJL89 plasmid.

Conflicts of Interest: The authors declare no conflict of interest.

References

1. Liu, S.-L.; Wang, Z.-G.; Xie, H.-Y.; Liu, A.-A.; Lamb, D.C.; Pang, D.-W. Single-virus tracking: From imaging methodologies to virological applications. *Chem. Rev.* **2020**, *120*, 1936–1979. [CrossRef] [PubMed]

2. Baulcombe, D.C.; Chapman, S.; Santa Cruz, S. Jellyfish green fluorescent protein as a reporter for virus infections. *Plant J.* **1995**, *7*, 1045–1053. [CrossRef] [PubMed]

3. Wei, Y.; Han, X.; Wang, Z.; Gu, Q.; Li, H.; Chen, L.; Sun, B.; Shi, Y. Development of a GFP expression vector for Cucurbit chlorotic yellows virus. *Virol. J.* **2018**, *15*, 1–4. [CrossRef] [PubMed]

4. Vasques, R.M.; Lacorte, C.; da Luz, L.L.; Aranda, M.A.; Nagata, T. Development of a new tobamovirus-based viral vector for protein expression in plants. *Mol. Biol. Rep.* **2019**, *46*, 97–103. [CrossRef]

5. Sandra, N.; Jailani, A.A.K.; Jain, R.K.; Mandal, B. Development of Soybean Yellow Mottle Mosaic Virus-Based Expression Vector for Heterologous Protein Expression in French Bean. *Mol. Biotechnol.* **2019**, *61*, 181–190. [CrossRef]

6. Ehrengruber, M.U.; Lundstrom, K.; Schweitzer, C.; Heuss, C.; Schlesinger, S.; Gähwiler, B.H. Recombinant Semliki Forest virus and Sindbis virus efficiently infect neurons in hippocampal slice cultures. *Proc. Natl. Acad. Sci. USA* **1999**, *96*, 7041–7046. [CrossRef]

7. Verbeek, M.; Dullemans, A.M.; Van den Heuvel, J.; Maris, P.C.; Van der Vlugt, R.A.A. Identification and characterisation of tomato torrado virus, a new plant picorna-like virus from tomato. *Arch. Virol.* **2007**, *152*, 881. [CrossRef]

8. Bally, J.; Jung, H.; Mortimer, C.; Naim, F.; Philips, J.G.; Hellens, R.; Bombarely, A.; Goodin, M.M.; Waterhouse, P.M. The rise and rise of *Nicotiana benthamiana*: A plant for all reasons. *Annu. Rev. Phytopathol.* **2018**, *56*, 405–426. [CrossRef]

9. Verbeek, M.; van Bekkum, P.J.; Dullemans, A.M.; van der Vlugt, R.A.A. Torradoviruses are transmitted in a semi-persistent and stylet-borne manner by three whitefly vectors. *Virus Res.* **2014**, *186*, 55–60. [CrossRef]

10. Budziszewska, M.; Obrepalska-Steplowska, A.; Wieczorek, P.; Pospieszny, H. The nucleotide sequence of a Polish isolate of Tomato torrado virus. *Virus Genes* **2008**, *37*. [CrossRef]

11. Ferriol, I.; Turina, M.; Zamora-Macorra, E.J.; Falk, B.W. RNA1-independent replication and GFP expression from Tomato marchitez virus isolate M cloned cDNA. *Phytopathology* **2016**, *106*, 500–509. [CrossRef] [PubMed]

12. Ferriol, I.; Vallino, M.; Ciuffo, M.; Nigg, J.C.; Zamora-Macorra, E.J.; Falk, B.W.; Turina, M. The Torradovirus-specific RNA2-ORF1 protein is necessary for plant systemic infection. *Mol. Plant Pathol.* **2018**, *19*, 1319–1331. [CrossRef] [PubMed]

13. Wieczorek, P.; Wrzesińska, B.; Frackowiak, P.; Przybylska, A.; Obrępalska-Stęplowska, A. Contribution of Tomato torrado virus Vp26 coat protein subunit to systemic necrosis induction and virus infectivity in *Solanum lycopersicum* 06 Biological Sciences 0607 Plant Biology. *Virol. J.* **2019**, *16*. [CrossRef]

14. Wieczorek, P.; Obrepalska-Steplowska, A. A single amino acid substitution in movement protein of tomato torrado virus influences ToTV infectivity in *Solanum lycopersicum*. *Virus Res.* **2016**, *213*, 32–36. [CrossRef] [PubMed]

15. Wieczorek, P.; Obrępalska-Stęplowska, A. The N-terminal fragment of the tomato torrado virus RNA1-encoded polyprotein induces a hypersensitive response (HR)-like reaction in *Nicotiana benthamiana*. *Arch. Virol.* **2016**, *161*. [CrossRef]

16. Zarzyńska-Nowak, A.; Ferriol, I.; Falk, B.W.; Borodynko-Filas, N.; Hasiów-Jaroszewska, B. Construction of *Agrobacterium tumefaciens*-mediated tomato black ring virus infectious cDNA clones. *Virus Res.* **2017**, *230*, 59–62. [CrossRef]

17. Matsumura, E.E.; Coletta-Filho, H.D.; Machado, M.A.; Nouri, S.; Falk, B.W. Rescue of Citrus sudden death-associated virus in *Nicotiana benthamiana* plants from cloned cDNA: Insights into mechanisms of expression of the three capsid proteins. *Mol. Plant Pathol.* **2019**, *20*, 611–625. [CrossRef]

18. Green, M.R.; Hughes, H.; Sambrook, J.; MacCallum, P. Molecular cloning: A laboratory manual. In *Molecular Cloning: A Laboratory Manual*; Cold Spring Harbor Laboratory Press: Cold Spring Harbor, NY, USA, 2012; p. 1890.

19. Wieczorek, P.; Budziszewska, M.; Obrępalska-Stęplowska, A. Construction of infectious clones of tomato torrado virus and their delivery by agroinfiltration. *Arch. Virol.* **2014**, *160*. [CrossRef]

20. Ferriol, I.; Silva Junior, D.M.; Nigg, J.C.; Zamora-Macorra, E.J.; Falk, B.W. Identification of the cleavage sites of the RNA2-encoded polyproteins for two members of the genus Torradovirus by N-terminal sequencing of the virion capsid proteins. *Virology* **2016**, *498*, 109–115. [CrossRef]

21. Budziszewska, M.; Pospieszny, H.; Obrępalska-Stęplowska, A. Genome Characteristics, Phylogeny and Varying Host Specificity of Polish Kra and Ros Isolates of Tomato torrado virus. *J. Phytopathol.* **2016**, *164*, 281–285. [CrossRef]

22. Wieczorek, P.; Obrepalska-Steplowska, A. Multiplex RT-PCR reaction for simultaneous detection of tomato torrado virus and pepino mosaic virus co-infecting *Solanum lycopersicum*. *J. Plant Prot. Res.* **2013**, *53*, 289–294. [CrossRef]

23. Gong, Y.-N.; Tang, R.-Q.; Zhang, Y.; Peng, J.; Xian, O.; Zhang, Z.-H.; Zhang, S.-B.; Zhang, D.-Y.; Liu, H.; Luo, X.-W. The NIa-Protease Protein Encoded by the Pepper Mottle Virus Is a Pathogenicity Determinant and Releases DNA Methylation of *Nicotiana benthamiana*. *Front. Microbiol.* **2020**, *11*, 102. [CrossRef] [PubMed]

24. Shi, X.; Cordero, T.; Garrigues, S.; Marcos, J.F.; Daròs, J.; Coca, M. Efficient production of antifungal proteins in plants using a new transient expression vector derived from tobacco mosaic virus. *Plant Biotechnol. J.* **2019**, *17*, 1069–1080. [CrossRef] [PubMed]

25. Ibrahim, A.; Odon, V.; Kormelink, R. Plant viruses in plant molecular pharming: Towards the use of enveloped viruses. *Front. Plant Sci.* **2019**, *10*, 803. [CrossRef]

26. Almazán, F.; González, J.M.; Pénzes, Z.; Izeta, A.; Calvo, E.; Plana-Durán, J.; Enjuanes, L. Engineering the largest RNA virus genome as an infectious bacterial artificial chromosome. *Proc. Natl. Acad. Sci. USA* **2000**, *97*, 5516–5521. [CrossRef] [PubMed]

27. Choi, I.-R.; French, R.; Hein, G.L.; Stenger, D.C. Fully biologically active in vitro transcripts of the eriophyid mite-transmitted wheat streak mosaic tritimovirus. *Phytopathology* **1999**, *89*, 1182–1185. [CrossRef]

28. Hellens, R.P.; Edwards, E.A.; Leyland, N.R.; Bean, S.; Mullineaux, P.M. pGreen: A versatile and flexible binary Ti vector for Agrobacterium-mediated plant transformation. *Plant Mol. Biol.* **2000**, *42*, 819–832. [CrossRef]

Publisher's Note: MDPI stays neutral with regard to jurisdictional claims in published maps and institutional affiliations.

Review

Plant-Based Vaccines: The Way Ahead?

Zacharie LeBlanc [1,*], **Peter Waterhouse** [1,2] **and Julia Bally** [1,*]

[1] Centre for Agriculture and the Bioeconomy, Queensland University of Technology (QUT),
    Brisbane, QLD 4000, Australia; peter.waterhouse@qut.edu.au

[2] ARC Centre of Excellence for Plant Success in Nature and Agriculture, Queensland University of
    Technology (QUT), Brisbane, QLD 4000, Australia

* Correspondence: leblancz@qut.edu.au (Z.L.); Julia.bally@qut.edu.au (J.B.)

Academic Editor: Jeanmarie Verchot
Received: 21 November 2020; Accepted: 19 December 2020; Published: 22 December 2020

Abstract: Severe virus outbreaks are occurring more often and spreading faster and further than ever. Preparedness plans based on lessons learned from past epidemics can guide behavioral and pharmacological interventions to contain and treat emergent diseases. Although conventional biologics production systems can meet the pharmaceutical needs of a community at homeostasis, the COVID-19 pandemic has created an abrupt rise in demand for vaccines and therapeutics that highlight the gaps in this supply chain's ability to quickly develop and produce biologics in emergency situations given a short lead time. Considering the projected requirements for COVID-19 vaccines and the necessity for expedited large scale manufacture the capabilities of current biologics production systems should be surveyed to determine their applicability to pandemic preparedness. Plant-based biologics production systems have progressed to a state of commercial viability in the past 30 years with the capacity for production of complex, glycosylated, "mammalian compatible" molecules in a system with comparatively low production costs, high scalability, and production flexibility. Continued research drives the expansion of plant virus-based tools for harnessing the full production capacity from the plant biomass in transient systems. Here, we present an overview of vaccine production systems with a focus on plant-based production systems and their potential role as "first responders" in emergency pandemic situations.

Keywords: biopharming; vaccines; viruses; viral vectors; *Nicotiana benthamiana*; COVID-19; plant-based biologics production

---

1. Introduction

Biopharming is the use of a living system as a host for the manufacture of non-natively produced, biologic drugs. Using living systems as bio-factories can allow for economical production of complex biologics at large scales that may not be possible or economically feasible with current in vitro synthesis technologies. The first instance of this practice was the use of the bacterial host, *Escherichia coli*, to produce insulin in 1978 by Genentech, which was later commercialized in 1982 [1]. This alleviated the need for harvesting insulin from natural biological sources such as dog and calf pancreases [2]. The next technological leap for biopharming was the adoption of eukaryotic cells as production hosts, which allowed for the production of more complex molecules with mammalian type post translational modifications. This technology was first commercialized by Genentech in 1987 by repurposing *E. coli* fermenters for Chinese Hamster Ovary (CHO) cell production of the anticoagulant Activase [3,4]. This technological development was a boon for commercialization and CHO cells were quickly adopted as the preferred large-scale production host for complex therapeutic molecules. In 2017, the monoclonal antibodies (mAb) market was valued at 123 billion USD with 87% of newly approved mAb products being produced in CHO cells [5]. Developments in CHO cell biologics production technology have generated an efficient platform that is regarded as the industry standard, commercially available kits advertise a human antibody titer of 3 g/L with some groups reporting titers of >5.8 g/L [6,7]. The widespread adoption of this technology platform has led to government agencies developing regulatory frameworks that narrowly fit cell suspension-based production systems [8], and has consequently created hurdles for technologies that do not fit this format. Eukaryotic cell suspensions can be considered the second iteration of biopharming technology following prokaryotic production systems; however, in the biopharming space, a universal biologics production system does not yet exist. There are alternative systems that could avoid the expensive fermentation infrastructure, complex culturing conditions and lengthy development timelines associated with CHO cells. Currently biologics are produced in bacterial, yeast, mammalian, avian, insect, and plant systems. Advantages and disadvantages of these systems have been extensively reviewed and continuous developments have increased the yield and quality of biologics to the benchmark set by mammalian production systems [9,10]. The biologics production space is mainly dominated by fermentation-based technologies which in many cases require a lead time of as much as 12 months to select clones, optimize culturing conditions, and reach production capacity [11]. Transgenic animals, whole plants and embryonated hen's eggs (EHE) stand apart as non-fermentation-based biologics productions hosts that have been used for commercial production. In this category, whole plants require the lowest input costs for biomass amplification and give the greatest production flexibility when used in transient expression systems. The coronavirus disease 2019 (COVID-19) pandemic has created an immediate need for vaccines and therapeutics to mitigate the spread and lethality of this disease. This urgent requirement for medicine has prompted our analysis of the current biologics production systems and their respective capacities for expedited drug development and scaling to large scale production. The goal of this review is to analyze and contrast the range of biopharming systems available, with particular emphasis on plant-based platforms, in the context of emergency pandemic response.

2. Plant Biopharming

2.1. Development of Biologics Production Systems in Plants

The first recorded example of biopharming in plants was the production of chimeric human growth hormone via transgenic tobacco and sunflower by Barta et al in 1986 [12]. The low infrastructure cost and simple biomass amplification requirements associated with plants compared to fermentation-based systems spurred an immense amount of interest in the possibilities of using plants as cheap biofactories and, by using the appropriate crop species, edible vaccines. This was soon followed by efforts to demonstrate the capacity for scaling plant-based biologics production in fields by using stably transformed crop plants such as maize, barley, safflower, and rice as production hosts. Although this approach held promise, early adopters of the technology were challenged by public perception of genetically modified plants, transgenic plant containment issues and a regulatory system with no precedent for good manufacturing practice (GMP) pharmaceuticals produced in this system [8]. In the following decades further investigation into plant-based production systems has been explored in a wide cross section of the plant kingdom including microalgae, moss, sundews, pitcher plants [13], melon [14], tomato, carrot, lettuce, tobacco, *Nicotiana benthamiana*, corn, rice, wheat, soybean, barley, and sunflower. The first generation of commercial biologics production in plants was centered on whole transgenic plants [15]. Today, this landscape is occupied by both transgenic and transient whole plant production systems as well as cell-culture-based systems and plant-based cell free systems [16]. The first genetically engineered plant derived therapeutic approved by the FDA was produced by Protalix in 2012. Protalix

Biotherapeutics of Israel uses a transgenic carrot cell suspension system to produce taliglucerase alfa, for treatment of Gaucher disease [17]. Their production system is bioreactor-based and claims to have lower initial investment and running costs compared to mammalian-based systems [18]. Though many plant systems have been investigated for biologics production the current mainstream production host choice is *Nicotiana benthamiana*. It is the core production host of many companies including Medicago (https://www.medicago.com), Kentucky BioProcessing (https://kentuckybioprocessing.com), PlantForm (https://www.plantformcorp.com), Icon Genetics (https://www.icongenetics.com/), iBio (https://www.ibioinc.com), CapeBio (https://capebiosa.com), Bioapp (http://bioapp.co.kr) and Leaf Expression Systems (https://www.leafexpressionsystems.com). *N. benthamiana* was embraced by the research community because of its high susceptibility to pathogens which made it an excellent system for the study of plant pathogen interactions [19]. This Australian native plant is thought to have adopted a life strategy of sacrificing pathogen defenses in favor of a hastened reproduction cycle. This remarkable susceptibility to infection, by viruses in particular, is thought to play a role in the plant's amenability to genetic transformation and high level transient gene expression, making it an excellent protein production host [20]. Transient expression in *N. benthamiana* allows the production of recombinant products in days rather than the 3- to 6-month timeline necessary when developing stable transgenic plants. In a typical *N. benthamiana* transient expression protocol, plants are grown to 4–6 weeks old then infected with a strain of the plant pathogen *Agrobacterium tumefaciens* containing genes of interest (GOI). *A. tumefaciens* transfers multiple copies of the GOI expression cassette to *N. benthamiana* which the plant then expresses in infected cells, the GOI product will typically reach peak level following a 5- to 7-day period. Product recovery can be achieved by homogenizing plant material and purification by a combination of filtration and chromatography methods. This system has been refined for biologics production by knocking out glycotransferases causing plant specific N linked glycosylations as well as development of methodologies for incorporation human type N and O glycosylations [21–23]. These refinements allow for production of recombinant proteins with mammalian glycosylation profiles. Further improvements to this system are continually arising with the goals of increasing product yield and quality by modifying the plant host, the *A. tumefaciens* strain, the infection methodologies and the DNA expression vector system [24]. In recent years, viral vector systems have provided the highest boosts in product yield in this transient system.

2.2. Viral Expression Vectors in Plants

A clear example of biopharming found in nature is the virus, which is an obligate parasite by definition, specializing in host invasion and redirection of biological processes for its own proliferation. Viral infection can commandeer host protein production systems causing accumulation of viral particles to 10% of plant dry weight [25]. This figure is likely the highest production of non-native protein in plants and is seen as the theoretical upper limit for transient protein production. Viral components have become a mainstay in plant biotechnology since the discovery of the cauliflower mosaic virus promoter in 1985, which was found to direct constitutive gene expression in most plant tissues and resulted in the highest known transgene expression at the time [26]. The use of viral components was further expanded by repurposing viral RNA silencing suppressors, such as P19 or V2, which overcome the RNA silencing machinery of the plant and inhibit degradation of foreign RNA [27,28]. When viral RNA silencing suppressors were expressed simultaneously with a GOI it resulted in a 50-fold increase of target protein yields [29]. A landmark discovery was the demonstration that GOIs could be inserted into the viral genome taking advantage of virus mobility and proliferation [30]. This "whole virus" approach is considered the first generation of viral vectors, whereby a GOI is inserted as a viral coat protein fusion or in place of the viral coat protein and relies on native virus infectivity and replication for GOI protein production [31]. The utility of this first approach was limited by non-comprehensive leaf coverage, low yields and insert size limitations, as viruses were shown to quickly lose the inserted gene during passage [30,32]. The second generation of viral vectors dubbed "deconstructed viral vectors" remove unnecessary viral component such as the coat protein while maintaining 5′ and 3′ UTR and replicase components. Deconstructed viral vectors rely on Agrobacterium infection for delivery to plants and the viral components for amplification and spread of the transgenic nucleic acid from cell to cell [25]. Notable examples of deconstructed viral vectors are the tobacco mosaic virus (TMV) derived magnICON and TRBO systems, the cowpea mosaic virus (CPMV) derived pEAQ and various potato virus X (PVX) based systems (Table 1). The first iteration of the magnICON system allowed for larger insert sizes, comprehensive tissue coverage and target product yields as high as 40% of total soluble protein or 4 g per kg fresh weight, with subsequent iterations of the technology reaching levels as high as 5.5 g/kg fresh weight [33,34]. Improvements in the most common systems based on TMV, CPMV, and PVX are typically achieved by removing and/or shuffling viral components and combining then into single vectors [34]. A common limitation of deconstructed viral vectors is their capacity for only one gene of interest per vector, which can be problematic for the expression of multichain products such as antibodies. This can be resolved by co-infiltration with non-competing TMV and PVX based systems [35]. The derivation of viral vectors from viruses is a field under constant development with the goals of expanding plant host range, increasing target protein yields, discovery of viral systems that can work in concert and mitigating deleterious effects to the production host. For example, foxtail mosaic virus has recently been shown to give improved monocot transformation, increased product yields and greater insert carrying capacity over the more traditional barley strip mosaic virus and wheat streak mosaic virus based systems [36]. Viral expression systems have cemented their position as a key component for high yielding transient expression and are likely to be the cornerstone of any commercialization venture involving biopharming in plants. Transient expression with viral vectors in *N. benthamiana* is a modular system with a flexibility not seen in other complex biologics production systems.

Table 1. Example of plant viruses used as viral expression vectors and their selected applications.

Virus	Genome	Production Host	System, Comment, Reference
Alfalfa mosaic virus (Alfamovirus)	I (+) ssRNA	*N. benthamiana*	VLPs/CP [37,38]
Bamboo mosaic virus (Potexvirus)	F (+) ssRNA	*N. benthamiana*, *Chenopodium. quinoa*	Full length viral vectors [39]
Beet Curly top virus (Curtovirus)	T-I (+) ssDNA	*N. benthamiana*	Deconstructed viral vectors [40–42]
Bean yellow dwarf virus (Mastrevirus)	T-I (+) ssDNA	*N. benthamiana*, *Nicotiana tabacum*, lettuce	Deconstructed viral vectors/VLPs [39,40,42,43]
Brome mosaic virus (Bromovirus)	I (+) ssRNA	Barley	1st plant RNA virus/VLPs [39,43]
Beet necrotic yellow vein virus (Benyvirus)	RS (+) ssRNA	*N. benthamiana*, *C. quinoa*	Deconstructed viral vectors [44]
Bean pod mottle virus (Comovirus)	I (+) ssRNA	Soybean, *P. sativum*	Deconstructed viral vectors [45]
Barley stripe mosaic virus (Hordeivirus)	RS (+) ssRNA	Black-grass	Deconstructed viral vectors [46,47]
Cauliflower mosaic virus (Caulimovirus)	I dsDNA	*Brassica rapa*	1st viral vector (Constitutive promoter)/Full length and deconstructed viral vectors [39]
Catharantus yellow mosaic virus (Begomovirus)	T I (+) ssDNA	*Catharanthus roseus*	Deconstructed viral vectors [47]

Table 1. *Cont.*

Virus	Genome	Production Host	System, Comment, Reference
Cucumber green mottle mosaic virus (Tobamovirus)	RS (+) ssRNA	Muskmelon	Full length viral vectors [48]
Cucumber mosaic virus (Cucumovirus)	I (+) ssRNA	*N. benthamiana*	Deconstructed viral vectors/VLPs [37,39,49]
Cowpea mosaic virus (Comovirus)	I (+) ssRNA	*Vigna unguiculata*	1st virus applied as an epitope presentation system/Full length and deconstructed viral vectors/VLPs [37,50]
Citrus tristeza virus (Closterovirus)	F (+) ssRNA	Citrus trees	Deconstructed viral vectors [25]
Foxtail mosaic virus (Potexvirus)	F (+) ssRNA	Maize, wheat, black-grass	Deconstructed viral vectors [36]
Hibiscus chlorotic ringspot virus (Betacarmovirus)	I (+) ssRNA	Kenaf leaves	VLPs [39]
Odontoglossum ringspot virus (Tobamovirus)	RS (+) ssRNA	*N. benthamiana*	Deconstructed viral vectors (hybrid with TMV) [25,39]
Papaya mosaic virus (Potexvirus)	RS (+) ssRNA	*E. coli*	VLPs [39]
Pea early browning virus (Tobravirus)	RS (+) ssRNA	*N. benthamiana*	Deconstructed viral vectors [36]
Pepper ringspot virus (Tobravirus)	RS (+) ssRNA	*N. benthamiana*	Deconstructed viral vectors [49]
Plum pox potyvirus (Potyvirus)	F R-S (+) ssRNA	*Nicotiana clevelandii*	Full length and deconstructed viral vectors [37,40,49,51]
Potato virus X (Potexvirus)	F (+) ssRNA	*N. benthamiana*	Full length and deconstructed viral vectors/VLPs [37,50]
Sun hemp mosaic virus (Tobamovirus)	RS (+) ssRNA	*N. benthamiana*, cowpea, lentil	Deconstructed viral vectors [52]
Tomato bushy stunt virus (Tombusvirus)	I (+) ssRNA	*N. benthamiana*, *Nicotiana excelsiana*	Deconstructed viral vectors [37,50]
Tobacco etch virus (Potyvirus)	RS (+) ssRNA	*Medicago trunculata*	Full length viral vectors [51]
Tomato golden mosaic virus (Begomovirus)	I I (+) ssDNA	*N. benthamiana*	Deconstructed viral vectors [39]
Tobacco mild green mosaic virus (Tobamovirus)	RS (+) ssRNA	*N. benthamiana*	Deconstructed viral vectors [52]
Tobacco mosaic virus (Tobamovirus)	RS (+) ssRNA	*N. benthamiana, N. excelsiana*	Full length and deconstructed viral vectors/VLPs [37,39,49]
Tomato mosaic virus (Tobamovirus)	RS (+) ssRNA	*N. tabacum*	Deconstructed viral vectors (hybrid with TMV) [39]
Triticum mosaic virus (Tritimovirus)	F (+) ssRNA	Wheat, maize	Deconstructed viral vectors [53]
Tobacco rattle virus (Tobravirus)	RS (+) ssRNA	*N. benthamiana*	Deconstructed viral vectors [54]
Turnip vein-clearing virus (Tobamovirus)	RS (+) ssRNA	*N. benthamiana*	Deconstructed viral vectors (hybrid with TMV) [39]
Tobacco yellow dwarf virus (Mastrevirus)	T I (I) ssDNA	*N. tabacum*	Deconstructed viral vectors [42]

Table 1. *Cont.*

Virus	Genome	Production Host	System, Comment, Reference
Wheat dwarf virus (Mastrevirus)	T I (+) ssDNA	*Triticum monococcum*	Deconstructed viral vectors [55]
Turnip yellow mosaic virus (Tymovirus)	I (+) ssRNA	Cabbage	VLps [39]
Wheat streak mosaic virus (Tritimovirus)	F (+) ssRNA	Wheat, maize	Deconstructed viral vectors [53]
Zucchini yellow mosaic virus (Potyvirus)	F R-S (+) ssRNA	Squash, melon cucumber	Deconstructed viral vectors (particle bombardment) [56]

I: Icosahedral, F: Filamentous, T: twinned, RS: rod-shaped, VLPs: Virus Like Particles, CP: Coat Protein.

3. Systems for Vaccine Manufacture

Viral outbreaks of the past decade have solidified the perspective that containment is best achieved by quick detection informing nonpharmaceutical interventions followed by vaccination [57]. During the influenza A (H1N1) pandemic of 2009, aside from issues related to vaccine sharing and proper distribution, one of the key failings was insufficient global vaccine production capacity and production speed, which was unable to mitigate the spread of the first wave of infection. This was primarily a result of reliance on egg-based vaccine manufacturing systems with slow production speeds [58]. In theory, with proper communication, virus spread can be halted primarily through testing, isolation, and contact tracing of infected individuals followed by vaccinations pre-empting viral transmission to new areas [59–61]. These strategies were not put into practice for the SARS-CoV-2 pandemic which has caused massive shut downs in many parts of the world and provided the public with an opportunity to learn about the duration of clinical trials for a vaccine candidate. The H1N1 and COVID-19 outbreaks have also highlighted gaps in the vaccine production pipeline. Since 1945, governments worldwide have been reliant on egg-based vaccine production which use EHE as a host to replicate viruses which are subsequently purified then inactivated or attenuated. While this system is proven and is still considered a primary failsafe for disease outbreaks, the drawbacks are obvious in the face of an outbreak requiring a reactive response. Production pipelines are limited by the quantity of fertilized eggs available; the subsequent processing requires 14 days and can provide 5–20 mg of virus per 100 eggs [62]. This is accomplished in the US by an annual investment of at least 57 million USD in farms at undisclosed locations under federal contract by the department of health and human services [63]. Viral amplification for vaccine production has also been ported to several mammalian cell lines including Madin-Darby Canine Kidney (MDCK) cells, Vero cells originating from African green monkey kidney cells, Medical Research Council cell strain 5 (MRC5) cells, and Wistar Institute WI-38 cells. Both MRC5 and WI-38 cells originate from human fetal lung tissue, which confers the advantage over EHE vaccines of having a reduced risk of vaccine inefficiency due to avian specific viral adaptation or virus selection during viral passage. Additionally, scalability is not bottlenecked by egg production [64,65]. A specific drawback for the use of MDCK and EHE based systems to respond to the COVID-19 pandemic is their inability to support the replication of SARS-CoV-1 and SARS-CoV-2 viruses [66]. This deficiency highlights the problem with relying on native viral amplification for vaccine production. Next generation vaccines are not made from natively amplified viruses but use specific recombinant viral peptides or virus like particles (VLPs), composed of viral structural proteins and/or membrane elements expressed and assembled in the production host. VLPs are structurally identical to wildtype viral particles but lack the genetic material required to replicate and, because they are not reliant on native virus infectivity for inoculation and amplification, they can be produced in a wider range of host organisms such as insect cell lines and plants. This production methodology offers the advantages of safety because no live virus is present during manufacture and there are greater scaling options due to the range of production hosts available. In many cases VLPs have been equivalent or superior in their ability to raise an immune response in mice as compared to live viruses [67]. The US Food and Drug Administration (FDA) currently has two VLP vaccines for protection against human papilloma virus. As of 24 April 2020, there are 97 vaccines licensed for use in the US by the FDA, 33 of which are derived from EHEs, 27 have components produced in mammalian cells, 5 contain components produced in yeast, and 3 contain components produced in insect cells (Table 2). For the 2019–2020 flu season the US will offer its first egg-free influenza vaccine. In Canada, Medicago Inc. has recently completed a phase 3 trial for a plant-made VLP quadrivalent flu vaccine, which is an important milestone for plant-made biologics [68,69]. As these next generation vaccines begin to penetrate the market, this new technology promises more precise protection as well as a wider range of production options.

Table 2. European Medicines Agency (EMA) and FDA Approved Vaccines 2020 [70,71].

Vaccine Indication and Number Approved by EMA; FDA	Production System(s) Associated with Vaccine	Type of Vaccines
Adenovirus Type 4 and Type 7 Vaccine, (0;1)	WI-38 human diploid cells	Live virus
Anthrax Vaccine (0;1)	*Bacillus anthracis*	Protective antigen protein from cell filtrates
BCG Vaccine (0;2)	*Mycobacterium bovis*	Attenuated bacteria
Cholera Vaccine (2;1)	*Vibrio cholera*	Attenuated bacteria
Dengue Vaccine (1;1)	Vero Cells	Live virus
Diphtheria and/or Tetanus and/or Acellular Pertussis and/or Hepatitis B and/or Polio and/or Hemophilus b and/or Vaccine (6;17)	*Corynebacterium diphtheriae, Clostridium tetani, Bordetella pertussis,* vero cells, *Saccharomyces cerevisiae, Haemophilus influenzae* type b and *Neisseria meningitidis*	Toxoids, antigens, inactivated virus, outer membrane protein, recombinant protein and capsular polysaccharide
Ebola Zaire Vaccine (3;1)	Vero cells	Live recombinant viral vaccine
Hepatitis A and/or Hepatitis B Vaccine (5;6)	MRC-5 human diploid cells, *Hansenula polymorpha* and *S. cerevisiae*	Inactivated virus, recombinant protein and VLP
Human Papillomavirus Vaccine (3;3)	*S. cerevisiae* and Baculovirus in *Trichoplusia ni* cells	VLP
Influenza A H1N1 Vaccine (1;7)	Embryonated chicken eggs	Inactivated virus and live virus
Influenza A H5N1 Vaccine (6;2)	Embryonated chicken eggs and MDCK cells	Inactivated virus
Influenza Vaccine (3;29)	Embryonated chicken eggs, MDCK cells and Sf9 cells	Inactivated virus and recombinant HA protein
Japanese Encephalitis Virus Vaccine (1;1)	Vero cells	Inactivated virus
Measles, Mumps, and Rubella Virus Vaccine (1;1)	Chick embryo cell culture and WI-38 human diploid cells	Attenuated and live virus vaccine
Measles, Mumps, Rubella and Varicella Virus Vaccine (1,1)	Chick embryo cell culture, WI-38 human diploid cells and MRC-5 human diploid cells	Attenuated and live virus vaccine

Table 2. *Cont.*

Vaccine Indication and Number Approved by EMA; FDA	Production System(s) Associated with Vaccine	Type of Vaccines
Meningococcal Vaccine (3;6)	*C. diphtheriae, N. meningitidis, E. coli* and *C. tetani*	Capsular polysaccharide, capsular polysaccharide toxoid conjugate, recombinant protein and outer membrane vesicle
Pneumococcal Vaccine (2;2)	*Streptococcus pneumoniae* and *C. diphtheriae*	Capsular polysaccharide and capsular polysaccharide protein conjugate
Rabies Vaccine (Human) (0;2)	MRC-5 human diploid cells and chicken fibroblasts	Inactivated virus
Rotavirus Vaccine (2;2)	Vero cells	Live attenuated virus
Smallpox (1;1)	Vero cells	Live virus
Smallpox and Monkeypox Vaccine (0;1)	Chicken embryo fibroblasts	Live virus
Typhoid Vaccine (2;1)	*Salmonella typhi* Ty21a	Live attenuated virus, capsular polysaccharide
Varicella Virus Vaccine (0;1)	WI-38 human diploid cells, MRC-5 human diploid cells	Live attenuated virus
Yellow Fever Vaccine (1;1)	Living avian leukosis virus-free chicken embryos	Attenuated virus
Zoster Vaccine (2;2)	MRC-5 human diploid cells, CHO and *Salmonella minnesota*	Live attenuated virus and virus surface glycoprotein E (gE) antigen c

Plant Systems for Viral Outbreak Response

Despite a modest presence of products on the pharmaceutical market, plant biopharming systems have been demonstrated on several occasions to be effective biologics production hosts, with the full capacity to produce correctly folded and glycosylated therapeutic molecules. In 2001, the Blue Angel Project sponsored by the US Defense Advanced Research Projects Agency sought to address, "insufficient capability to provide vaccines against pandemics caused by new strains, as well as infections caused by intentional biothreats", by demonstrating the vaccine production capabilities of plant-based systems, by 1. developing a hardened, high containment, self-sufficient plant-based pharmaceutical production facility; 2. building a facility with the capacity to manufacture 10 million doses of an H1N1 influenza vaccine in a single month; and 3. completing this project within an 18 month window [72]. This project demonstrated that the plant-based production systems were capable of quick vaccine production and have the production pace that would be required to quell an unexpected viral outbreak. It was successfully completed in different stages by Medicago Inc., Caliber Biotherapeutics Inc. (now iBio Inc.), Fraunhofer CMB and Kentucky BioProcessing Inc. These companies operate currently as producers of biologics with portfolios including various vaccines and/or antibodies for cancer therapies. Today, Medicago reports that it can deliver mass quantities of a novel flu vaccine in a three-month timeline [73]. In 2014, The production speed of this system was demonstrated when Kentucky BioProcessing was able to quickly produce an Ebola antibody cocktail called Zmapp, developed by Mapp Biopharmaceutical, that had been granted emergency compassionate approval for human use [74]. This product, which is administered at 50 mg/kg, was produced in sufficient quantities to be used for the treatment of six people infected with

Ebola, five of whom recovered. More recently, Medicago was able to produce VLP vaccine candidates 20 days after having access to the COVID-19 S protein sequence [75]. Although the long duration of clinical trials cannot be avoided, emergency governmental authorization to overlap clinical trials can shorten time to deployment for vaccines; making vaccine development and production timelines the bottlenecks prolonging the time to deployment [76].

The ability of plant-based biologics production facilities to quickly shift production pipelines for emergency manufacturing runs could be a great asset for pandemic situations and should be considered as an added value of these facilities by government sponsors. As seen in previous emergency pandemic responses, nations with vaccine production capabilities have had difficulties distributing vaccines to other countries without incentive [58]. This lack of vaccine sharing is likely to be repeated in the COVID-19 pandemic considering the economic consequences that this pandemic has brought already. These situations exemplify the need for decentralized biologics production lines to provide security for coming pandemics. With a comparatively low infrastructure cost, estimated at <50% of the cost of fermentation-based systems, plant-based biologics platforms make local vaccine and therapeutic production a more attainable goal for countries currently lacking pharmaceutical industry [77].

4. Conclusions and Future Perspectives

The immediate need for biologics in response to COVID-19 and the perceived lack of infrastructure to fill this demand has prompted analysis, by several groups, of how plant-based production systems can fill this need [78–81]. The ability of plant-based production systems to quickly pivot production to a variety of different target molecules and quickly produce large quantities at low cost is advantageous for pandemic response [73]. Indeed, several plant-based biologics manufacturers, using transient *N. benthamiana* expression systems, have initiated production of COVID-19-related products. Medicago and Kentucky BioProcessing have vaccines in clinical trial stages and iBio has 2 vaccine candidates and a therapeutic product currently in pre-clinical development. CapeBio and PlantForm Corporation are currently developing kits for COVID-19 testing and Leaf Expression Systems is producing viral proteins to support COVID-19 research and development. These corporations should be commended for their reactivity to the situation, and perhaps signal that plant-based platforms are now sufficiently developed to be a mainstream part of the plan to combat future outbreaks.

At the time of writing, there are 48 COVID-19 vaccines in the human trial phase, worldwide, with projected public release in early to mid-2021. Assuming a typical flu dosage of 45 ug per person, 45 g of vaccine will be required for 1 million people, which would scale to 351 kg for the world population. In addition to the need for vaccinations, there is also a requirement for therapeutic antibodies for those infected with COVID-19. In a recent review, Tusé et al. [78] estimated that the capacity of all mammalian cell fermentation facilities, worldwide, would be able to fulfill only 50% of this demand in one year, not including development time and assuming a low dose therapy (1 g per person). Current projections for vaccine release discuss prioritization of population segments for initial distribution, indicating a foreseen limitation in supplies. The COVID-19 pandemic will be resolved through a combined effort of different production systems to manufacture vaccines for public immunization as well as therapies for those infected. This situation has provided an opportunity to evaluate pandemic response systems globally and will be looked to in coming years for insight on the design of systems that can adequately respond to future outbreaks that are sure to come.

Funding: This work was funded by the Australian Research Council (ARC), grant number FL16010 0155 and DP170103960.

Conflicts of Interest: The authors declare no conflict of interest.

References

1. Quianzon, C.C.; Cheikh, I. History of insulin. *J. Community Hosp. Intern. Med. Perspect.* **2012**, *2*. [CrossRef] [PubMed]
2. Chance, R.E.; Frank, B.H. Research, development, production, and safety of biosynthetic human insulin. *Diabetes Care* **1993**, *16* (Suppl. 3), 133–142. [CrossRef]
3. Papini, M.; Salazar, M.; Nielsen, J. Systems Biology of Industrial Microorganisms. In *Biosystems Engineering I: Creating Superior Biocatalysts*; Wittmann, C., Krull, R., Eds.; Springer: Berlin/Heidelberg, Germany, 2010; pp. 51–99.
4. Jayapal, K.P.; Wlaschin, K.F.; Hu, W.; Yap, M. Recombinant protein therapeutics from CHO cells: 20 years and counting. *Chem. Eng. Prog.* **2007**, *103*, 40–47.
5. Walsh, G. Biopharmaceutical benchmarks 2018. *Nat. Biotechnol.* **2018**, *36*, 1136–1145. [CrossRef] [PubMed]
6. Reinhart, D.; Damjanovic, L.; Kaisermayer, C.; Kunert, R. Benchmarking of commercially available CHO cell culture media for antibody production. *Appl. Microbiol. Biotechnol.* **2015**, *99*, 4645–4657. [CrossRef]
7. Thermo Fischer Scientific. ExpiCHO Expression System. Available online: https://www.thermofisher.com/au/en/home/life-science/protein-biology/protein-expression/mammalian-protein-expression/transient-mammalian-protein-expression/expicho-expression-system.html (accessed on 4 October 2020).
8. Fischer, R.; Buyel, J.F. Molecular farming—The slope of enlightenment. *Biotechnol. Adv.* **2020**, *40*, 107519. [CrossRef]
9. Legastelois, I.; Buffin, S.; Peubez, I.; Mignon, C.; Sodoyer, R.; Werle, B. Non-conventional expression systems for the production of vaccine proteins and immunotherapeutic molecules. *Hum Vaccines Immunother.* **2017**, *13*, 947–961. [CrossRef]
10. Shanmugaraj, B.; Malla, A.; Phoolcharoen, W. Emergence of Novel Coronavirus 2019-nCoV: Need for Rapid Vaccine and Biologics Development. *Pathogens* **2020**, *9*, 148. [CrossRef]
11. Lai, T.; Yang, Y.; Ng, S.K. Advances in Mammalian cell line development technologies for recombinant protein production. *Pharmaceuticals* **2013**, *6*, 579–603. [CrossRef]
12. Barta, A.; Sommergruber, K.; Thompson, D.; Hartmuth, K.; Matzke, M.A.; Matzke, A.J.M. The expression of a nopaline synthase—Human growth hormone chimaeric gene in transformed tobacco and sunflower callus tissue. *Plant Mol. Biol.* **1986**, *6*, 347–357. [CrossRef]
13. Miguel, S.; Nisse, E.; Biteau, F.; Rottloff, S.; Mignard, B.; Gontier, E.; Hehn, A.; Bourgaud, F. Assessing Carnivorous Plants for the Production of Recombinant Proteins. *Front. Plant Sci.* **2019**, *10*, 1–13. [CrossRef] [PubMed]
14. Kim, S.-R.; Sim, J.-S.; Ajjappala, H.; Kim, Y.-H.; Hahn, B.-S. Expression and large-scale production of the biochemically active human tissue-plasminogen activator in hairy roots of Oriental melon (Cucumis melo). *J. Biosci. Bioeng.* **2012**, *113*, 106–111. [CrossRef] [PubMed]
15. Hood, E.E. From green plants to industrial enzymes. *Enzym. Microb. Technol.* **2002**, *30*, 279–283. [CrossRef]
16. Buntru, M.; Vogel, S.; Spiegel, H.; Schillberg, S. Tobacco BY-2 cell-free lysate: An alternative and highly-productive plant-based in vitro translation system. *BMC Biotechnol.* **2014**, *14*, 37. [CrossRef] [PubMed]
17. Zimran, A.; Brill-Almon, E.; Chertkoff, R.; Petakov, M.; Blanco-Favela, F.; Muñoz, E.T.; Solorio-Meza, S.E.; Amato, D.; Duran, G.; Giona, F.; et al. Pivotal trial with plant cell-expressed recombinant glucocerebrosidase, taliglucerase alfa, a novel enzyme replacement therapy for Gaucher disease. *Blood* **2011**, *118*, 5767–5773. [CrossRef] [PubMed]
18. Protalix Biotherapeutics. Procellex Platform. Available online: http://protalix.com/technology/procellex-platform/ (accessed on 4 October 2020).
19. Goodin, M.M.; Zaitlin, D.; Naidu, R.A.; Lommel, S.A. *Nicotiana benthamiana*: Its history and future as a model for plant-pathogen interactions. *Mol. Plant Microbe Interact.* **2008**, *21*, 1015–1026. [CrossRef]
20. Bally, J.; Nakasugi, K.; Jia, F.; Jung, H.; Ho, S.Y.; Wong, M.; Paul, C.M.; Naim, F.; Wood, C.C.; Crowhurst, R.N.; et al. The extremophile *Nicotiana benthamiana* has traded viral defence for early vigour. *Nat. Plants* **2015**, *1*, 15165. [CrossRef]
21. Jansing, J.; Sack, M.; Augustine, S.M.; Fischer, R.; Bortesi, L. CRISPR/Cas9-mediated knockout of six glycosyltransferase genes in *Nicotiana benthamiana* for the production of recombinant proteins lacking β-1,2-xylose and core α-1,3-fucose. *Plant Biotechnol. J.* **2019**, *17*, 350–361. [CrossRef]

22. Kallolimath, S.; Castilho, A.; Strasser, R.; Grünwald-Gruber, C.; Altmann, F.; Strubl, S.; Galuska, C.E.; Zlatina, K.; Galuska, S.P.; Werner, S.; et al. Engineering of complex protein sialylation in plants. *Proc. Natl. Acad. Sci. USA* **2016**, *113*, 9498–9503. [CrossRef]

23. Castilho, A.; Neumann, L.; Daskalova, S.; Mason, H.S.; Steinkellner, H.; Altmann, F.; Strasser, R. Engineering of sialylated mucin-type O-glycosylation in plants. *J. Biol. Chem.* **2012**, *287*, 36518–36526. [CrossRef]

24. Sainsbury, F. Innovation in plant-based transient protein expression for infectious disease prevention and preparedness. *Curr. Opin. Biotechnol.* **2020**, *61*, 110–115. [CrossRef] [PubMed]

25. Dawson, W.O. A personal history of virus-based vector construction. *Curr. Top. Microbiol. Immunol.* **2014**, *375*, 1–18. [CrossRef] [PubMed]

26. Odell, J.T.; Nagy, F.; Chua, N.-H. Identification of DNA sequences required for activity of the cauliflower mosaic virus 35S promoter. *Nature* **1985**, *313*, 810–812. [CrossRef] [PubMed]

27. Naim, F.; Nakasugi, K.; Crowhurst, R.N.; Hilario, E.; Zwart, A.B.; Hellens, R.P.; Taylor, J.M.; Waterhouse, P.M.; Wood, C.C. Advanced Engineering of Lipid Metabolism in *Nicotiana benthamiana* Using a Draft Genome and the V2 Viral Silencing-Suppressor Protein. *PLoS ONE* **2012**, *7*, e52717. [CrossRef] [PubMed]

28. Csorba, T.; Kontra, L.; Burgyán, J. Viral silencing suppressors: Tools forged to fine-tune host-pathogen coexistence. *Virology* **2015**, *479–480*, 85–103. [CrossRef] [PubMed]

29. Voinnet, O.; Rivas, S.; Mestre, P.; Baulcombe, D. An enhanced transient expression system in plants based on suppression of gene silencing by the p19 protein of tomato bushy stunt virus. *Plant J.* **2003**, *33*, 949–956. [CrossRef]

30. Gronenborn, B.; Gardner, R.C.; Schaefer, S.; Shepherd, R.J. Propagation of foreign DNA in plants using cauliflower mosaic virus as vector. *Nature* **1981**, *294*, 773–776. [CrossRef]

31. Gleba, Y.; Klimyuk, V.; Marillonnet, S. Viral vectors for the expression of proteins in plants. *Curr. Opin. Biotechnol.* **2007**, *18*, 134–141. [CrossRef]

32. Dawson, W.O.; Lewandowski, D.J.; Hilf, M.E.; Bubrick, P.; Raffo, A.J.; Shaw, J.J.; Grantham, G.L.; Desjardins, P.R. A tobacco mosaic virus-hybrid expresses and loses an added gene. *Virology* **1989**, *172*, 285–292. [CrossRef]

33. Marillonnet, S.; Thoeringer, C.; Kandzia, R.; Klimyuk, V.; Gleba, Y. Systemic *Agrobacterium tumefaciens*-mediated transfection of viral replicons for efficient transient expression in plants. *Nat. Biotechnol.* **2005**, *23*, 718–723. [CrossRef]

34. Lindbo, J.A. TRBO: A high-efficiency tobacco mosaic virus RNA-based overexpression vector. *Plant Physiol.* **2007**, *145*, 1232–1240. [CrossRef] [PubMed]

35. Giritch, A.; Marillonnet, S.; Engler, C.; van Eldik, G.; Botterman, J.; Klimyuk, V.; Gleba, Y. Rapid high-yield expression of full-size IgG antibodies in plants coinfected with noncompeting viral vectors. *Proc. Natl. Acad. Sci. USA* **2006**, *103*, 14701–14706. [CrossRef] [PubMed]

36. Bouton, C.; King, R.C.; Chen, H.; Azhakanandam, K.; Bieri, S.; Hammond-Kosack, K.E.; Kanyuka, K. Foxtail mosaic virus: A Viral Vector for Protein Expression in Cereals. *Plant Physiol.* **2018**, *177*, 1352–1367. [CrossRef]

37. Lico, C.; Chen, Q.; Santi, L. Viral vectors for production of recombinant proteins in plants. *J. Cell. Physiol.* **2008**, *216*, 366–377. [CrossRef] [PubMed]

38. Yusibov, V.; Modelska, A.; Steplewski, K.; Agadjanyan, M.; Weiner, D.; Hooper, D.C.; Koprowski, H. Antigens produced in plants by infection with chimeric plant viruses immunize against rabies virus and HIV-1. *Proc. Natl. Acad. Sci. USA* **1997**, *94*, 5784–5788. [CrossRef]

39. Ibrahim, A.; Odon, V.; Kormelink, R. Plant Viruses in Plant Molecular Pharming: Toward the Use of Enveloped Viruses. *Front. Plant Sci.* **2019**, *10*, 803. [CrossRef]

40. Hefferon, K. Plant Virus Expression Vectors: A Powerhouse for Global Health. *Biomedicines* **2017**, *5*, 44. [CrossRef]

41. Chung, H.Y.; Lee, H.H.; Kim, K.I.; Chung, H.Y.; Hwang-Bo, J.; Park, J.H.; Sunter, G.; Kim, J.B.; Shon, D.H.; Kim, W.; et al. Expression of a recombinant chimeric protein of hepatitis A virus VP1-Fc using a replicating vector based on Beet curly top virus in tobacco leaves and its immunogenicity in mice. *Plant Cell Rep.* **2011**, *30*, 1513–1521. [CrossRef]

42. Hefferon, K.L. DNA Virus Vectors for Vaccine Production in Plants: Spotlight on Geminiviruses. *Vaccines* **2014**, *2*, 642–653. [CrossRef]

43. French, R.; Janda, M.; Ahlquist, P. Bacterial gene inserted in an engineered RNA virus: Efficient expression in monocotyledonous plant cells. *Science* **1986**, *231*, 1294–1297. [CrossRef]

44. Jiang, N.; Zhang, C.; Liu, J.Y.; Guo, Z.H.; Zhang, Z.Y.; Han, C.G.; Wang, Y. Development of Beet necrotic yellow vein virus-based vectors for multiple-gene expression and guide RNA delivery in plant genome editing. *Plant Biotechnol. J.* **2019**, *17*, 1302–1315. [CrossRef] [PubMed]

45. Zhang, C.; Bradshaw, J.D.; Whitham, S.A.; Hill, J.H. The development of an efficient multipurpose bean pod mottle virus viral vector set for foreign gene expression and RNA silencing. *Plant Physiol.* **2010**, *153*, 52–65. [CrossRef] [PubMed]

46. Mellado-Sánchez, M.; McDiarmid, F.; Cardoso, V.; Kanyuka, K.; MacGregor, D.R. Virus-Mediated Transient Expression Techniques Enable Gene Function Studies in Black-Grass. *Plant Physiol.* **2020**, *183*, 455–459. [CrossRef] [PubMed]

47. Mortimer, C.; Dugdale, B.; Waterhouse, P. Development of an autonomously replicating viral expression system tailored for *Catharanthus roseus*. *Plant Biotechnol. J.* **2020**, *18*, 1115–1117. [CrossRef]

48. Ooi, A.; Tan, S.; Mohamed, R.; Rahman, N.A.; Othman, R.Y. The full-length clone of cucumber green mottle mosaic virus and its application as an expression system for Hepatitis B surface antigen. *J. Biotechnol.* **2006**, *121*, 471–481. [CrossRef]

49. Tavares-Esashika, M.L.; Campos, R.N.S.; Blawid, R.; da Luz, L.L.; Inoue-Nagata, A.K.; Nagata, T. Characterization of an infectious clone of pepper ringspot virus and its use as a viral vector. *Arch. Virol.* **2020**, *165*, 367–375. [CrossRef]

50. Nagyová, A.; Subr, Z. Infectious full-length clones of plant viruses and their use for construction of viral vectors. *Acta Virol.* **2007**, *51*, 223–237.

51. Dolja, V.V.; McBride, H.J.; Carrington, J.C. Tagging of plant potyvirus replication and movement by insertion of beta-glucuronidase into the viral polyprotein. *Proc. Natl. Acad. Sci. USA* **1992**, *89*, 10208–10212. [CrossRef]

52. Liu, Z.; Kearney, C.M. A tobamovirus expression vector for agroinfection of legumes and *Nicotiana*. *J. Biotechnol.* **2010**, *147*, 151–159. [CrossRef]

53. Yang, N.; Peng, C.; Cheng, D.; Huang, Q.; Xu, G.; Gao, F.; Chen, L. The over-expression of calmodulin from Antarctic notothenioid fish increases cold tolerance in tobacco. *Gene* **2013**, *521*, 32–37. [CrossRef]

54. Peyret, H.; Lomonossoff, G.P. When plant virology met *Agrobacterium*: The rise of the deconstructed clones. *Plant Biotechnol. J.* **2015**, *13*, 1121–1135. [CrossRef] [PubMed]

55. Matzeit, V.; Schaefer, S.; Kammann, M.; Schalk, H.J.; Schell, J.; Gronenborn, B. Wheat dwarf virus vectors replicate and express foreign genes in cells of monocotyledonous plants. *Plant Cell* **1991**, *3*, 247–258. [CrossRef] [PubMed]

56. Arazi, T.; Lee Huang, P.; Huang, P.L.; Zhang, L.; Moshe Shiboleth, Y.; Gal-On, A.; Lee-Huang, S. Production of antiviral and antitumor proteins MAP30 and GAP31 in cucurbits using the plant virus vector ZYMV-AGII. *Biochem. Biophys. Res. Commun.* **2002**, *292*, 441–448. [CrossRef] [PubMed]

57. WHO Ebola Response Team. After Ebola in West Africa—Unpredictable Risks, Preventable Epidemics. *N. Engl. J. Med.* **2016**, *375*, 587–596. [CrossRef] [PubMed]

58. Fineberg, H.V. Pandemic preparedness and response—Lessons from the H1N1 influenza of 2009. *N. Engl. J. Med.* **2014**, *370*, 1335–1342. [CrossRef] [PubMed]

59. Blaisdell, L.L.; Cohn, W.; Pavell, J.R.; Rubin, D.S.; Vergales, J.E. Preventing and Mitigating SARS-CoV-2 Transmission—Four Overnight Camps, Maine, June–August 2020. *MMWR Morb. Mortal. Wkly. Rep.* **2020**, *69*. [CrossRef] [PubMed]

60. Jernigan, D.B. Update: Public Health Response to the Coronavirus Disease 2019 Outbreak—United States. *MMWR Morb. Mortal. Wkly. Rep.* **2020**, *69*. [CrossRef]

61. Choi, W.; Shim, E. Optimal strategies for vaccination and social distancing in a game-theoretic epidemiologic model. *J. Theor. Biol.* **2020**, *505*, 110422. [CrossRef]

62. Ping, J.; Lopes, T.J.S.; Nidom, C.A.; Ghedin, E.; Macken, C.A.; Fitch, A.; Imai, M.; Maher, E.A.; Neumann, G.; Kawaoka, Y. Development of high-yield influenza A virus vaccine viruses. *Nat. Commun.* **2015**, *6*, 8148. [CrossRef]

63. Hendrickson, J.R. The Coronavirus and Lessons for Preparedness. *Spec. Ed. Policy Brief.* **2020**, *2020*. [CrossRef]

64. Minor, P.D.; Engelhardt, O.G.; Wood, J.M.; Robertson, J.S.; Blayer, S.; Colegate, T.; Fabry, L.; Heldens, J.G.M.; Kino, Y.; Kistner, O.; et al. Current challenges in implementing cell-derived influenza vaccines: Implications for production and regulation, July 2007, NIBSC, Potters Bar, UK. *Vaccine* **2009**, *27*, 2907–2913. [CrossRef] [PubMed]

65. Manini, I.; Trombetta, C.M.; Lazzeri, G.; Pozzi, T.; Rossi, S.; Montomoli, E. Egg-Independent Influenza Vaccines and Vaccine Candidates. *Vaccines* **2017**, *5*, 18. [CrossRef] [PubMed]

66. Barr, I.G.; Rynehart, C.; Whitney, P.; Druce, J. SARS-CoV-2 does not replicate in embryonated hen's eggs or in MDCK cell lines. *Eurosurveillance* **2020**, *25*. [CrossRef] [PubMed]

67. Bright, R.A.; Carter, D.M.; Daniluk, S.; Toapanta, F.R.; Ahmad, A.; Gavrilov, V.; Massare, M.; Pushko, P.; Mytle, N.; Rowe, T.; et al. Influenza virus-like particles elicit broader immune responses than whole virion inactivated influenza virus or recombinant hemagglutinin. *Vaccine* **2007**, *25*, 3871–3878. [CrossRef]

68. Centers for Disease Control and Prevention. Recombinant Influenza (Flu) Vaccine. Available online: https://www.cdc.gov/flu/prevent/qa_flublok-vaccine.htm (accessed on 4 October 2020).

69. Ward, B.J.; Makarkov, A.; Séguin, A.; Pillet, S.; Trépanier, S.; Dhaliwall, J.; Libman, M.D.; Vesikari, T.; Landry, N. Efficacy, immunogenicity, and safety of a plant-derived, quadrivalent, virus-like particle influenza vaccine in adults (18–64 years) and older adults (≥65 years): Two multicentre, randomised phase 3 trials. *Lancet* **2020**, *396*, 1491–1503. [CrossRef]

70. FDA. Vaccines Licensed for Use in the United States. Available online: https://www.fda.gov/vaccines-blood-biologics/vaccines/vaccines-licensed-use-united-states (accessed on 4 October 2020).

71. European Medicines Agency (EMA). Download Medicine Data. Available online: https://www.ema.europa.eu/en/medicines/download-medicine-data (accessed on 9 December 2020).

72. Holtz, B.R.; Berquist, B.R.; Bennett, L.D.; Kommineni, V.J.M.; Munigunti, R.K.; White, E.L.; Wilkerson, D.C.; Wong, K.-Y.I.; Ly, L.H.; Marcel, S. Commercial-scale biotherapeutics manufacturing facility for plant-made pharmaceuticals. *Plant Biotechnol. J.* **2015**, *13*, 1180–1190. [CrossRef]

73. Greer, A.L. Early vaccine availability represents an important public health advance for the control of pandemic influenza. *BMC Res. Notes* **2015**, *8*, 191. [CrossRef]

74. Zhang, Y.; Li, D.; Jin, X.; Huang, Z. Fighting Ebola with ZMapp: Spotlight on plant-made antibody. *Sci. China Life Sci.* **2014**, *57*, 987–988. [CrossRef]

75. Medicago. Medicago COVID-19 Vaccine Development Program. Available online: https://www.medicago.com/en/covid-19-programs/ (accessed on 4 October 2020).

76. US Department of Defense. Coronavirus: Operation Warp Speed. Available online: https://www.defense.gov/Explore/Spotlight/Coronavirus/Operation-Warp-Speed/ (accessed on 4 October 2020).

77. Nandi, S.; Kwong, A.T.; Holtz, B.R.; Erwin, R.L.; Marcel, S.; McDonald, K.A. Techno economic analysis of a transient plant-based platform for monoclonal antibody production. *MAbs* **2016**, *8*, 1456–1466. [CrossRef]

78. Tusé, D.; Nandi, S.; McDonald, K.A.; Buyel, J.F. The Emergency Response Capacity of Plant-Based Biopharmaceutical Manufacturing-What It Is and What It Could Be. *Front. Plant Sci.* **2020**, *11*. [CrossRef]

79. Capell, T.; Twyman, R.M.; Armario-Najera, V.; Ma, J.K.; Schillberg, S.; Christou, P. Potential Applications of Plant Biotechnology against SARS-CoV-2. *Trends Plant Sci.* **2020**, *25*, 635–643. [CrossRef] [PubMed]

80. Rosales-Mendoza, S. Will plant-made biopharmaceuticals play a role in the fight against COVID-19? *Expert Opin. Biol. Ther.* **2020**, *20*, 545–548. [CrossRef] [PubMed]

81. McDonald, K.A.; Holtz, R.B. From Farm to Finger Prick—A Perspective on How Plants Can Help in the Fight Against COVID-19. *Front. Bioeng. Biotechnol.* **2020**, *8*, 782. [CrossRef] [PubMed]

Publisher's Note: MDPI stays neutral with regard to jurisdictional claims in published maps and institutional affiliations.

MDPI

St. Alban-Anlage 66

4052 Basel

Switzerland

Tel. +41 61 683 77 34

Fax +41 61 302 89 18

www.mdpi.com

Viruses Editorial Office

E-mail: viruses@mdpi.com

www.mdpi.com/journal/viruses